Lecture Notes in Mathematics

1943

Editors:
J.-M. Morel, Cachan
F. Takens, Groningen
B. Teissier, Paris

FONDAZIONE CIME
ROBERTO CONTI

CENTRO INTERNAZIONALE MATEMATICO ESTIVO
INTERNATIONAL MATHEMATICAL SUMMER CENTER

C.I.M.E. means Centro Internazionale Matematico Estivo, that is, International Mathematical Summer Center. Conceived in the early fifties, it was born in 1954 ánd made welcome by the world mathematical community where it remains in good health and spirit. Many mathematicians from all over the world have been involved in a way or another in C.I.M.E.'s activities during the past years.

So they already know what the C.I.M.E. is all about. For the benefit of future potential users and co-operators the main purposes and the functioning of the Centre may be summarized as follows: every year, during the summer, Sessions (three or four as a rule) on different themes from pure and applied mathematics are offered by application to mathematicians from all countries. Each session is generally based on three or four main courses ($24-30$ hours over a period of $6-8$ working days) held from specialists of international renown, plus a certain number of seminars.

A C.I.M.E. Session, therefore, is neither a Symposium, nor just a School, but maybe a blend of both. The aim is that of bringing to the attention of younger researchers the origins, later developments, and perspectives of some branch of live mathematics.

The topics of the courses are generally of international resonance and the participation of the courses cover the expertise of different countries and continents. Such combination, gave an excellent opportunity to young participants to be acquainted with the most advance research in the topics of the courses and the possibility of an interchange with the world famous specialists. The full immersion atmosphere of the courses and the daily exchange among participants are a first building brick in the edifice of international collaboration in mathematical research.

C.I.M.E. Director
Pietro ZECCA
Dipartimento di Energetica "S. Stecco"
Università di Firenze
Via S. Marta, 3
50139 Florence
Italy
e-mail: zecca@unifi.it

C.I.M.E. Secretary
Elvira MASCOLO
Dipartimento di Matematica
Università di Firenze
viale G.B. Morgagni 67/A
50134 Florence
Italy
e-mail: mascolo@math.unifi.it

For more information see CIME's homepage: http://www.cime.unifi.it

CIME's activity is supported by:

- Ministero degli Affari Esteri, Direzione Generale per la Promozione e la Cooperazione, Ufficio V
- Consiglio Nazionale delle Ricerche
- E.U. under the Training and Mobility of Researchers Programme

Luis L. Bonilla (Ed.)

Inverse Problems and Imaging

Lectures given at the
C.I.M.E. Summer School
held in Martina Franca, Italy
September 15–21, 2002

Chapters by:
A. Carpio · O. Dorn · M. Moscoso · F. Natterer
G.C. Papanicolaou · M.L. Rapún · A. Teta

 Springer

FONDAZIONE
CIME
ROBERTO CONTI

Luis L. Bonilla (Ed.)

Gregorio Millán Institute of Fluid Dynamics
Nanoscience and Industrial Mathematics
Universidad Carlos III de Madrid
Avda. de la Universidad 30
28911 Leganés, Madrid, Spain
bonilla@ing.uc3m.es
http://scala.uc3m.es

Ana Carpio

Departamento de Matemática Aplicada
Facultad de Matemáticas
Universidad Complutense de Madrid
Plaza de Ciencias 3
28040 Madrid, Spain
ana_carpio@mat.ucm.es
www.mat.ucm.es/~acarpio

Oliver Dorn
Miguel Moscoso

Gregorio Millán Institute of Fluid Dynamics
Nanoscience and Industrial Mathematics
Universidad Carlos III de Madrid
Avda. de la Universidad 30
28911 Leganés, Madrid, Spain
moscoso@math.uc3m.es
odorn@math.uc3m.es

Frank Natterer

Institut für Numerische und Angewandte
Mathematik, University of Münster
Einsteinstraße 62
48149 Münster
Germany
nattere@math.uni-muenster.de

George C. Papanicolaou

Mathematics Department
Building 380, 383V
Stanford University
Stanford, CA 94305, USA
papanicolaou@stanford.edu

Maria Luisa Rapún

Departamento de Fundamentos
Matemáticos de la Tecnología Aeronáutica
Escuela Técnica Superior de Ingenieros
Aeronáuticos, Universidad Politécnica
de Madrid, Plaza del Cardenal Cisneros 3
28040 Madrid, Spain
marialuisa.rapun@upm.es

Alessandro Teta

Dipartimento di Matematica Pura
e Applicata, Università di L'Aquila
via Vetoio – loc. Coppito
67100 L'Aquila, Italy
teta@univaq.it

ISBN: 978-3-540-78545-3   e-ISBN: 978-3-540-78547-7
DOI: 10.1007/978-3-540-78547-7

Lecture Notes in Mathematics ISSN print edition: 0075-8434
                    ISSN electronic edition: 1617-9692

Library of Congress Control Number: 2008922185

Mathematics Subject Classification (2000): 65R32, 44A12, 92C55, 85A25, 35Q40

Cover design: *WMX Design* GmbH, Heidelberg

Printed on acid-free paper

9 8 7 6 5 4 3 2 1

springer.com

Preface

Nowadays, we are facing numerous important imaging problems, as for example, the detection of anti-personal land mines in post-war remediation areas, detection of unexploded ordnances (UXO), nondestructive testing of materials, monitoring of industrial processes, enhancement of oil production by efficient reservoir characterization, and the various exciting and emerging developments in noninvasive imaging techniques for medical purposes – computerized tomography (CT), magnetic resonance imaging (MRI), positron emission tomography (PET) and ultrasound tomography, to mention only a few. It is broadly recognized that these problems can only be solved by a joint effort of experts in mathematical and technical sciences.

The CIME Summer School on Imaging, held in Martina Franca, Italy, from 15 to 21 September, 2002, encompassed the theory and applications of imaging in different disciplines including, medicine, geophysics, engineering, etc. The Summer School brought together leading experts in mathematical techniques and applications in many different fields, to present a broad and useful introduction for non-experts and practitioners alike to many aspects of this exciting field. The main lecturers were Simon Arridge, Frank Natterer, George C. Papanicolaou and William Symes. Seminars on related special topics were contributed by Oliver Dorn, Miguel Moscoso and Alessandro Teta. Among the different topics dealt with in the school, we may cite X-ray tomography, diffusive optical tomography with possible applications to tumor detection in medicine by optical means, electromagnetic induction tomography used in geophysics, and techniques of seismic tomography to image the wave velocity structure of a region and to obtain information on possible oil or gas reservoirs, etc. Furthermore, there were extensive discussions on the mathematical bases for analyzing these methods. The mathematical and computational techniques that are used for imaging have many common features and form a rapidly developing part of applied mathematics.

The present volume contains a general introduction on image reconstruction by M. Moscoso, some of the lectures and presentations given in the Summer School (F. Natterer, O. Dorn et al., M. Moscoso, G. Dell'Antonio et al.), and two additional lectures on other imaging techniques by A. Carpio and M.L. Rapún and by O. Dorn.

The lectures by Prof. Frank Natterer introduce the mathematical theory and the reconstruction algorithms of computerized X-ray tomography. These lectures give a short account of integral geometry and the Radon transform, reconstruction algorithms such as the filtered back projection algorithm, iterative methods (for example, the Kaczmarz method) and Fourier methods. They also comment on the three-dimensional case, which is the subject of current research. Many of the fundamental tools and issues of computerized tomography, such as back projection, sampling, and high frequency analysis, have their counterparts in more advanced imaging techniques for impedance, optical or ultrasound tomography and are most easily studied in the framework of computerized tomography.

The chapter by O. Dorn, H. Bertete-Aguirre and G.C. Papanicolaou reviews electromagnetic induction tomography, used to solve imaging problems in geophysical and environmental imaging applications. The focus is on realistic 3D situations which provide serious computational challenges as well as interesting novel mathematical problems to the practitioners. The chapter first introduces the reader to the mathematical formulation of the underlying inverse problem; it then describes the theory of sensitivity analysis in this application; it proposes a nonlinear reconstruction algorithm for solving such problems efficiently; it discusses a regularization technique for stabilizing the reconstruction; and finally it presents various numerical examples for illustrating the discussed concepts and ideas.

The chapter by M. Moscoso presents optical imaging of biological tissue using the polarization effects of a narrow beam of light. The biological tissue is modeled as a continuous medium which varies randomly in space and which contains inhomogeneities with no sharp boundaries. This differs from the more usual point of view in which the biological tissue is modeled as a medium containing discrete spherical particles of the same or different sizes. The propagation of light is then described by a vector radiative transport equation which is solved by a Monte Carlo method. A discussion on how to use polarization to improve image reconstruction is given as well.

The chapter by A. Carpio and M.L. Rapún explains how to use topological derivative methods to solve constrained optimization reformulations of inverse scattering problems. This chapter gives formulas to calculate the topological derivatives for the Helmholtz equation and for the equations of elastic waves. Furthermore they explain and implement a practical iterative numerical scheme to detect objects based on computing the topological derivative of a cost functional associated to these equations in successive approximate domains. Many examples of reconstruction of objects illustrate this method.

The chapter by O. Dorn deals with an inverse problem in underwater acoustic and wireless communication. He establishes a link between the time-reversal and adjoint methods for imaging and proposes a method for solving the inverse problem based on iterative time-reversal experiments.

Lastly, the chapter by G. Dell'Antonio, R. Figari and A. Teta reviews the theory of Hamiltonians with point interactions, i.e., with potentials supported on a finite set of points. This chapter studies the mathematical basis of scattering with point scatterers, analyzes how such an idealized situation relates to short-range potentials and discusses situations in which the strength of the potential depends on the wave function as it has been proposed in the physics literature on double barriers and other nanostructures. Knowing the solution of the direct problem is of course a prerequisite to be able to image the scatterers from measurements on a boundary.

Universidad Carlos III de Madrid   *Luis L. Bonilla*

Contents

Introduction to Image Reconstruction
Miguel Moscoso.. 1
1 Introduction 1
2 Applications 3
   2.1 Medicine 3
   2.2 Geophysics 8
   2.3 Industry 10
3 Stability 12
4 Image Reconstruction Techniques..................................... 14
   4.1 Computerized Tomography 14
   4.2 Diffuse Optical Tomography 15
   4.3 Other Problems 16
References 16

X-ray Tomography
Frank Natterer... 17
1 Introduction 17
2 Integral Geometry 18
   2.1 The Radon Transform 18
   2.2 The Ray Transform.. 21
   2.3 The Cone-Beam Transform 22
   2.4 The Attenuated Radon Transform 23
   2.5 Vectorial Transforms.. 25
3 Reconstruction Algorithms 26
   3.1 The Filtered Backprojection Algorithm 26
   3.2 Iterative Algorithms 28
   3.3 Fourier Methods 31
4 Reconstruction from Cone Beam Data 32
References 33

Adjoint Fields and Sensitivities for 3D Electromagnetic Imaging in Isotropic and Anisotropic Media
Oliver Dorn, Hugo Bertete-Aguirre, and George C. Papanicolaou 35
1 Introduction: Electromagnetic Imaging of the Earth 35
2 Sensitivities and the Adjoint Method 38
3 Maxwell's Equations for Anisotropic Media 41
4 The Linearized Residual Operator 43
5 Adjoint Sensitivity Analysis for Anisotropic Media................... 45
    5.1 The Adjoint Linearized Residual Operator..................... 45
    5.2 Decomposition of the Residual Operator and Its Adjoint 46
    5.3 A Special Case: Isotropic Media............................. 47
6 A Single Step Adjoint Field Inversion Scheme 48
    6.1 Outline of the Inversion Scheme 48
    6.2 Efficient Calculation of the Adjoint of the Linearized
          Residual Operator.. 51
    6.3 Iterative Algorithm for the Adjoint Field Inversion 52
7 Regularization and Smoothing with Function Spaces 53
8 Numerical Examples for Electromagnetic Sensitivity Functions 56
    8.1 Some Numerically Calculated Sensitivity Functions 56
    8.2 A Short Discussion of Sensitivity Functions 60
References 63

Polarization-Based Optical Imaging
Miguel Moscoso.. 67
1 Introduction 67
2 Radiative Transfer Equation 68
3 Model 70
4 Monte Carlo Method 71
    4.1 Analog Monte Carlo Sampling 72
    4.2 Splitting and Russian Roulette.............................. 74
    4.3 Point Detectors 75
5 Numerical Results 77
6 Conclusions... 81
References 82

Topological Derivatives for Shape Reconstruction
Ana Carpio and Maria Luisa Rapún............................... 85
1 Introduction 85
2 Helmholtz Equations... 87
    2.1 The Forward and Inverse Scattering Problems 87
    2.2 Topological Derivatives Applied to Inverse Scattering 89
3 Numerical Computation of Topological Derivatives 92
4 An Iterative Method Based on Topological Derivatives 101
5 Numerical Solution of Forward and Adjoint Problems 107
6 Explicit Expressions for the Topological Derivatives................. 112
    6.1 Shape Derivative 112
    6.2 Proof of Theorems 2.1 and 2.2 118

7 Sounding Solids by Elastic Waves 122
   7.1 The Forward and Inverse Scattering Problems 122
   7.2 Shape Derivatives 125
   7.3 Topological Derivative 129
8 Conclusions.. 131
References 131

Time-Reversal and the Adjoint Imaging Method
with an Application in Telecommunication
Oliver Dorn 135
1 Introduction 135
2 Time-Reversal and Communication in the Ocean 137
3 The MIMO Setup in Wireless Communication 139
4 Symmetric Hyperbolic Systems................................. 140
5 Examples for Symmetric Hyperbolic Systems 141
   5.1 Linearized Acoustic Equations 142
   5.2 Maxwell's Equations 142
   5.3 Elastic Waves Equations 143
6 The Basic Spaces and Operators 144
7 An Inverse Problem Arising in Communication.................... 146
8 The Basic Approach for Solving the Inverse Problem
   of Communication 148
9 The Adjoint Operator A^*...................................... 149
10 The Acoustic Time-Reversal Mirror............................. 150
11 The Electromagnetic Time-Reversal Mirror 152
12 Time-Reversal and the Adjoint Operator A^* 155
13 Iterative Time-Reversal for the Gradient Method 156
   13.1 The Basic Version 156
   13.2 The Regularized Version 158
14 The Minimum Norm Solution Approach 158
   14.1 The Basic Version 158
   14.2 The Regularized Version 160
15 The Regularized Least Squares Solution Revisited 161
16 Partial and Generalized Measurements 162
17 Summary and Future Research Directions 164
References 168

A Brief Review on Point Interactions
Gianfausto Dell'Antonio, Rodolfo Figari, and Alessandro Teta 171
1 Introduction 171
2 Construction of Point Interactions 173
3 Connection with Smooth Interactions 182
4 Time Dependent Point Interactions 185
References 188

List of Participants 191

Introduction to Image Reconstruction

Miguel Moscoso

Gregorio Millán Institute of Fluid Dynamics, Nanoscience and Industrial
Mathematics, Universidad Carlos III de Madrid, Leganés, Spain
moscoso@math.uc3m.es

Summary. The problem of reconstructing images from measurements at the boundary of a domain belong to the class of inverse problems. In practice, these measurements are incomplete and inaccurate leading to ill-posed problems. This means that 'exact' reconstructions are usually not possible. In this Introduction the reader will find some applications in which the main ideas about stability and resolution in image reconstruction are discussed. We will see that although different applications or imaging modalities work under different physical principles and map different physical parameters, they all share the same mathematical foundations and the tools used to create the images have a great deal in common. Current imaging problems deal with understanding the trade off between data size, the quality of the image and the computational tools used to create the image. In many cases, these tools represent the performance bottleneck due to the high operational count and the memory cost.

1 Introduction

Imaging is a broad field which covers all aspects of the analysis, modification, compression, visualization, and generation of images. It is a highly interdisciplinary field in which researchers from biology, medicine, engineering, computer science, physics, and mathematics, among others, work together to provide the best possible image. Imaging science is profoundly mathematical and challenging from the modeling and the scientific computing point of view [1–5].

There are at least two major areas in imaging science in which applied mathematics has a strong impact: image processing, and image reconstruction. In image processing the input is a (digital) image such as a photograph, while in image reconstruction the input is a set of data. In the latter case, the data is limited, and its poor information content is not enough to generate an image to start with.

Image processing and analysis of information in images are methods that become increasingly important in many technical and scientific fields. Image processing techniques treat an image and apply numerical algorithms to either improve the given image or to extract different features of it. Image segmentation is typically used for the latter purpose. It refers to the process of partitioning an image into multiple regions (locating objects and boundaries) in order to simplify its representation for its further analysis. Each region shares the same properties or characteristics such as color, intensity or texture. Many different techniques have been applied for image segmentation. We mention here, graph partitioning methods in which the image is modeled as a graph, level-sets methods in which an initial shape is evolved towards the object boundary, or statistical methods in which we view a region of an image as one realization of the various random processes involved in the formation of that region (probability distribution functions and histograms are used to estimate the characteristics of the regions). We will not discuss the mathematics of image processing here.

Image reconstruction refers to the techniques used to create an image of the interior of a body (or region) non-invasively, from data collected on its boundary [1–5]. Image reconstruction can be seen as the solution of a mathematical inverse problem in which the cause is inferred from the effect. As a consequence, measurement and recording techniques designed to produce the images depend deeply on the application we consider. In medicine, for example, very different procedures are applied depending on the disease the physician is looking for. Among these imaging techniques there are now in routine use, computerized tomography, magnetic resonance, ultrasound, positron emission tomography (PET), or electroencephalography (EEG), among others. They all acquire data susceptible to be represented as maps containing positional information of medical interest but, while each procedure operates on a different physical background principle and is sensitive to a different characteristic of the human body (X-ray attenuation, dielectric properties, hydrogen nucleus density, reflection by tissue interfaces, ...), they all share the same mathematical formulation of the problem and similar numerical algorithms that allow the reconstruction of the desired image.

We want to stress here, though, that it is always necessary a deep understanding of a mathematical model with which to interpret the measurements and design numerical algorithms to reconstruct the desired images. Progress can hardly be carried out without efficient and well designed numerical solvers. Imaging is more than showing that an inverse problem may have a unique solution under circumstances that are rarely satisfied. Modern imaging approaches deal with understanding the trade off between data size, the quality of the image, the computational complexity of the forward model used to generate the measurements, and the complexity and stability of the numerical algorithm employed to obtain the images. One neither has all the data he wants, nor can solve a very general forward model to invert the data.

The purpose of this contribution is two fold: to give a first insight into different applications in which imaging has led to an important development, and to outline the close mathematical connections between the imaging techniques used in these applications. In Sect. 2 we review the impact that imaging has had in medicine, geophysics and industry. In Sect. 3 we discuss the main ideas about stability and resolution in imaging to introduce the reader to the basic notions and difficulties encountered in ill-posed inverse problems. Finally, in Sect. 4, we present different image reconstruction techniques in computerized tomography (which can be viewed as a linear inverse problem) and in diffuse optical tomography (which can be viewed as a nonlinear inverse problem).

2 Applications

The problem of reconstructing images from boundary data is of fundamental importance in many areas of science, engineering and industry. We will briefly describe applications in different areas in which *imaging* has seen a significant growth and innovation over the last decades. The presentation below has been organized in three subsections: medicine, geophysics, and industry applications. This organization is, however, somehow artificial. Seismic imaging, for example, is described within the Geophysics subsection, but it is clear that seismic imaging is of fundamental importance in the oil industry. Many of the imaging techniques presented in each subsection have been used along the years, in one form or another, within different applications as well. Magnetic resonance imaging, for example, is described in the Medicine subsection since it is primarily used to produce high quality images of the inside of the human body. However, magnetic resonance is also used (in a different setting), for example, in oil exploration and production since it can provide good estimates of permeability and fluid flow.

2.1 Medicine

X-ray tomography, ultrasound or magnetic resonance imaging have dramatically changed the practice in medicine [2–5]. These techniques are now in routine use. Nevertheless, they are still under current improvement, and an important part of the scientific research is focused towards the limits these techniques can reach. At the same time, new imaging techniques have been proposed along the last decades with some advantages over the more traditional techniques. These new techniques are the subject of a lot of academic research and their benefits need to be investigated thoroughly in order to assess their validity as daily diagnostic techniques. The development of these new technologies requires the combine effort of many people with very distinct specialization.

Next, we mention four imaging techniques. The first two are well established and are of daily use in our hospitals. The last two are still under development and might have an enormous medical impact in a near future.

X-ray Tomography

One of the main revolutions which medicine has experienced during the last few decades began in 1972 with the first clinical application of X-ray tomography or computerized tomography [2–5]. However, the theoretical foundations underlying these image reconstruction technique are due to the mathematician Johan Radon, back in 1917.

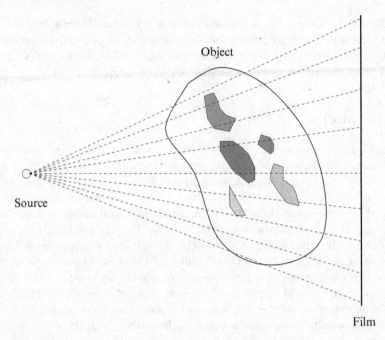

Fig. 1. An X-ray source illuminates a 2D object from a single view. Measurements, as line integrals of the attenuation coefficient over each line, are registered on the film plane. Observe that depth information is lost

X-ray tomography development was due to the necessity of overcoming some limitations of the X-ray images as they were used in their time. The main drawback of a traditional X-ray image is that it represents a three-dimensional object in a two-dimensional film. X-rays go through the human body, and some fraction of the incident radiation is absorbed (or scattered) by the internal structures of the body. The overall attenuation of the incident X-ray is proportional to the average density of the body along the straight line joining the X-ray source with a point of the film, in which this information is registered (Fig. 1). Since these measurements involve line integrals the third dimension is lost. This means that different structures that are completely separate in the body may overlap in the final image. This is particularly troublesome in traditional X-ray images that are read *by eye* by the physician.

The first experiments on medical applications were carried out by the physicist Alan Cormack between 1957 and 1964. He derived a quantitative theory for calculating the attenuation distribution of X-rays inside of the human body from the acknowledgment of the transmitted X-rays. He actually learned of Radon's work much later regretting the time lost. The first practical implementation of X-ray tomography was achieved by the engineer Godfrey Hounsfield in 1972. Like Cormack, Hounsfield realized that three-dimensional information could be retrieved from several two-dimensional projections, and like him, he worked without the knowledge of previous work. As a historical note, we mention that the first applications of Radon's theory were done by Bracewell in 1956 in the context of radioastronomy, but his work had little impact and was not known in the medical world. Both, Cormack and Hounsfield were awarded the Nobel Prize for medicine in 1979. The new and powerful computers needed for the implementation of the algorithms, were crucial for the success of this technology.

A description of the theory of imaging in X-ray tomography can be found in the lecture by Frank Natterer in this book. He surveys integral geometry and related imaging techniques. Computerized tomography reconstruction algorithms, such as the filtered backprojection algorithm, Kaczmarz's method, or Fourier methods, are also presented.

Magnetic Resonance Imaging

While X-ray tomography uses ionizing radiation to form images, magnetic resonance imaging (MRI) uses non-ionizing signals. Magnetic resonance was originally named nuclear magnetic resonance, but the word *nuclear* was dropped because of the negative connotations associated with the word nuclear. Neither people nor local politicians wanted anything related to *nuclear* in their communities. The term nuclear simply referred to the fact that all atoms have a nucleus. The measurements are taken under powerful magnetic fields in conjunction with radio waves. When placed in an external magnetic field, particles with spin different than zero, such as protons, can absorb energy that is later reemitted producing secondary radio waves. The detected signal results from the difference between the absorbed and the emitted energies.

Magnetic resonance imaging, proposed by Richard Ernst in 1975, is probably the major competitor to X-ray tomography. In 1980 Edelstein and coworkers demonstrated imaging of the body with his technique. Richard Ernst was rewarded for his contributions to the development of high resolution techniques in magnetic resonance imaging with the Nobel Prize in Chemistry. At the beginning of the 90s magnetic resonance imaging started to be used for functional imaging purposes. This new application allows the study of the function of the various regions of the human brain. In 2003, Paul C. Lauterbur and Sir Peter Mansfield were awarded the Nobel Prize in Medicine for their seminal discoveries concerning the use of magnetic resonance to visualize different structures which led to revolutionary insights into the functions of the brain and the workings of the human body.

Besides its ability for functional imaging, magnetic resonance imaging is also better suited for soft tissue than X-ray tomography. It can image different tissue properties by variation of the scanning parameters, and tissue contrast can be changed and enhanced in various ways to detect different features. The disadvantages of magnetic resonance imaging over X-ray tomography is that it is more expensive and more time consuming. Nevertheless, the imaging techniques behind this modality are the same used to reconstruct images in X-ray tomography.

Diffuse Optical Tomography

Can we use light to *see* or image inside the body? Diffuse optical tomography in the near-infrared is an emerging modality with potential in medical diagnosis and biomedicine. A near-infrared light source impinges upon the surface and the light propagates through the body (Fig. 2). The light that emerges is detected at an array and used to infer the local optical properties (absorption and scattering) of the illuminated tissue. The major absorbers when using near-infrared light are water and both oxygenated and deoxygenated hemoglobin. Cancer metastasis requires the growth of a new network of blood vessels (angiogenesis). The greater supply of blood around the tumors give rise to an absorbing obstacle feasible to be detected by this technique.

Another important current application of diffuse optical tomography is brain imaging. By shining near-infrared light on the scalp, changes in neural activity can be measured because of the tight correlation existing between brain activation and the changes in concentration of oxygenated and deoxygenated hemoglobin (cerebral hemodynamics). This is possible due to the relationship of the absorption spectra of oxygenated hemoglobin and deoxygenated hemoglobin at near-infrared wavelengths. In addition, diffuse optical tomography can detect and localize important events such as hemorrhagic strokes in the brain.

We would like to emphasize that to reconstruct quantitative images of the absorption and scattering parameters from near-infrared light measurements one

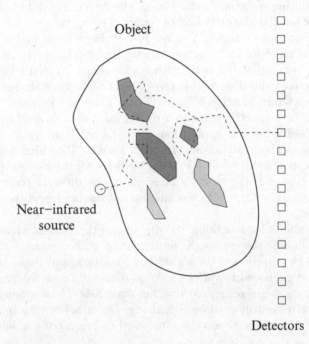

Fig. 2. A near-infrared source illuminates a 2D object. The photons do not travel along straight lines. This causes blurring

needs a mathematical model to interpret the data. The most comprehensive model for diffuse optical tomography is the radiative transport equation

$$\frac{1}{v}\frac{\partial I(\mathbf{x},\hat{\mathbf{k}},t)}{\partial t} + \hat{\mathbf{k}} \cdot \nabla_{\mathbf{x}} I(\mathbf{x},\hat{\mathbf{k}},t) + (\Sigma^{(s)}(\mathbf{x}) + \Sigma^{(a)}(\mathbf{x})) I(\mathbf{x},\hat{\mathbf{k}},t) =$$
$$\Sigma^{(s)}(\mathbf{x}) \int P(\hat{\mathbf{k}} \cdot \hat{\mathbf{k}}') I(\mathbf{x},\hat{\mathbf{k}}',t)\, d\hat{\mathbf{k}}' \, , \tag{1}$$

where v, $\Sigma^{(s)}$, and $\Sigma^{(a)}$ are the velocity of the light in the medium, the scattering coefficient, and the absorption coefficient, respectively. The fundamental quantity of radiative transfer is the specific intensity I which depends on the position vector $\mathbf{x}$ and unit direction vector $\hat{\mathbf{k}}$. The phase function $P(\hat{\mathbf{k}} \cdot \hat{\mathbf{k}}')$ in (1) describes the directional distribution of light that scatters in direction $\hat{\mathbf{k}}$ due to light of unit energy density in direction $\hat{\mathbf{k}}'$. We have considered here the scalar radiative transport equation.

The radiative transport equation (1) takes into account absorption and scattering due to the inhomogeneities in tissue. Analytical solutions to this integrodifferential equation are known only for relatively simple problems. For more complicated problems numerical solutions are needed. The use of this theory for imaging purposes in biological tissue can be found in the lecture by Miguel Moscoso. A Monte Carlo method to solve this equation can also be found.

Before we proceed, we mention that one can use the diffusion equation to model light propagation in tissue as an alternative to the more complicated radiative transport equation. The diffusion equation is much easier to solve but has limitations such that it does not describe accurately light propagation in optically thin media nor light near sources and boundaries. Nevertheless, the diffusion approximation to the radiative transport equation is widely used in the literature.

Microwave Breast Imaging

Microwave imaging is being investigated as a possible alternative to mammography. It uses electromagnetic data in the range from 300 MHz to 3 GHZ. Due to the large contrast that exists between normal breast tissue and malignant tumors, microwave imaging is also showing its potential to complement the mammography to differentiate between malignant tumors and other possible lesions that may appear in the breast. The key property of this new imaging modality is that the contrast between malignancies and healthy tissue at microwave frequencies is significantly greater than the small contrast exploited by X-ray mammography.

This technique is still the subject of a lot of research from the engineering and mathematical point of views. Regarding the mathematical point of view, a major limitation is the lack of efficient reconstruction algorithms that are able to produce sharp images of small tumors. We recall here, that the large contrast exhibited by the tumor give rise to a highly non-linear inverse problem. Therefore, one-step method like the Born and Rytov approximations are not well suited for this application. On the other hand, in this application one is not so much interested on the detailed reconstruction of the spatial distribution of the dielectric properties, but to know, in a reliable way, whether or not there is a tumor, its location and size, and it malignancy.

To answer these questions, different coherent systems, that measure phases (or arrival times) as well as intensities, are under investigation. (We remember here that X-ray tomography and diffuse optical tomography are incoherent imaging modalities that only use intensities as input.) Other reconstruction algorithms, based on the evolution of initial shapes that model the tumor boundary are also under development. Shape-based approaches offer several advantages, as for example, well defined boundaries and the incorporation of a-priori assumptions regarding the general anatomical structures in the body that help to simplify, and thereby, to stabilize the reconstruction algorithm. For the evolution law of the initial shape that model the tumor, sensitivity distributions of the associated inverse problem are usually used.

2.2 Geophysics

Traditionally, geophysics is an interdisciplinary science in which modern applied mathematics have had a crucial role in quantitative modeling, analysis, and computational development. All of these issues are essential to interpret and understand the measurements, as well as to asses the validity of the modeling. Once again, modeling and scientific computing are the first steps to develop efficient and robust imaging tools.

We now outline two applications of interest for imaging earth's interior. They involve different techniques depending on the nature and depth of the targets to image. Seismic imaging uses information carried by elastic waves, and reflections appear at boundaries with different impedances, while ground penetrating radar uses electromagnetic waves and reflections appear at boundaries with different dielectric constants.

Seismic Imaging

Seismic surveying involves sending sound waves underground and recording the reflected signals from the earth's interior [1]. Reflection occurs when the waves hit a change in medium and experience a change in their velocity. This is a simple fact. But changes in the earth's properties take place in very different spatial scales and are not easy to model. As a first approximation the earth is a stratified medium in which the thickness of the layer might vary from less than a meter to several hundreds of meters. On the other hand, tectonic forces in the earth can bend these layers or can give rise to fractures. Besides, the interior structure can be severely affected by the influence of other processes like the extreme heat generated from the earth's interior at some locations. As a result, the seismic waves propagating through the earth will be refracted, reflected, and diffracted from the heterogeneous medium, and the recorded signal will show a very complicated structure. Earth scientists use these data to create a three-dimensional model of the subsurface, i.e., a spatial image of a property of the earth's interior (typically the impedance or the wave velocity).

In order to accomplish this task, one must solve an inverse scattering problem to *undo* all these wave propagation effects and find the discontinuities. Migration or back-propagation methods are the dominant imaging tools used for focusing seismic images of subsurface structures. Seismic migration is used to transform seismic data into seismic images that can be interpreted as geologic sections. This technique uses an estimation of the local velocity to migrate or back-propagate the surface data down into the earth. Thereby, seismic reflection images are *moved* to their original position. Seismic migration is therefore, in some sense, the inverse of seismic modeling that describes the forward process of propagating waves from the sources to the scatterers, and to the receivers.

Migration of the seismic data is the most computationally expensive step in order to obtain the images. This process is usually accomplished by the Kirchhoff integral in which each measured data is summed into the image at the points where the measured travel time agrees with the total propagation time (derived from the assumed local velocity). The reflectors in the image are, therefore, built up by constructive interference of the data, collapsing energy from reflections back to their proper positions.

Electromagnetic Imaging

Electromagnetic data is a useful source of information for shallow subsurface earth characterization. Ground penetrating radar is a non-destructive technique that uses radar pulses to locate buried objects such as voids, contaminant plumes such as tank-leaks, pipelines, or cables in the earth's interior. It is also used to locate, map and study buried archaeological features.

Ground penetrating radar sends a pulse from the surface. The probing electromagnetic waves are usually between the radio and microwave range of frequencies. When the waves hit a buried object or a boundary with different dielectric constants, the reflected signals are received at the surface. The mathematical tools used to locate the object are similar to reflection seismology. The only difference is that electromagnetic waves are used instead of elastic waves. It is important to note that while seismic imaging can be adapted to any penetration depth, ground penetrating radar suffers an important limitation due to dissipation. The depth range depends on the electrical conductivity of the ground and the transmitted frequency. Typical penetration depths range from 5 to 10 m if the electrical conductivity of the medium is low.

Another popular electromagnetic method of shallow subsurface imaging is cross-borehole electromagnetic induction tomography. The source is a low frequency magnetic field (1 KHz) generated in one vertical borehole, and the signals are received at a more distant borehole. The goal of this modality is to image electrical conductivity variations in the earth. This technique can provide high resolution images between wells up to 1,000 m apart. Optimum operating frequencies, that depend on the borehole separation and ground conductivity value, ranges between 40 and 100 KHz. Lower frequencies limit the resolution but have a larger penetration range. Higher frequencies provide better resolution but limit the range of the measurements. The chapter by Oliver Dorn, Hugo Bertete-Aguirre, and George Papanicolaou presents this modality of imaging and a method for solving the inverse Maxwell problem in three dimensions.

More recent method involve deployment of both vertical borehole arrays and large surface arrays.

2.3 Industry

It is apparent that imaging has transformed the development of many industries. This is the case of some of the applications mentioned above such as seismic imaging which is the basis of crucial exploration, development, and production decisions in the oil industry.

We now survey two other industrial applications in which new imaging tools specifically designed for industrial purposes have been developed.

Nondestructive Testing

The problem in which one needs to detect the presence of fissures or flaws from measurements on the surface of a material is of much interest in industry. Nondestructive testing is any examination procedure performed on an object (without *destroying* it) in order to detect the presence of defects such as cracks that can alter its usefulness [3]. Nondestructive testing has become, in fact, an integral part of almost all industrial processes in which a product failure might result in an accident or body injury. For example, in aircraft maintenance it is important to inspect the mechanical damage and to evaluate the amount of work necessary to repair it. It is also important to do it in an efficient way, as the maintenance of aircraft must be accomplished within a scheduled time. Nondestructive testing can accomplish the inspection and can detect cracks or any other irregularities in the airframe structure that are not visible. In other words, nondestructive testing is the technology which provides information regarding the condition of the material under examination once the necessary measurements have been performed.

It is difficult to say when or who performed the first nondestructive testing. Nondestructive tests can actually be performed in a very natural way without employing sophisticated technologies. Listening to the sound of different metals has been used along the centuries to determine the quality of the piece being shaped (a sword, a ring, a bell, ...). However, it was in the late 1950s when nondestructive techniques started to experience an important growth through the development of new technologies and ideas. The mathematical study of this inverse problem was initiated by A. Friedman and M. Vogelius in 1989.

There is a broad range of techniques that are applied nowadays depending on the kind of measured data and the specific industrial situation under consideration. Boundary data can be obtained from thermal, acoustic, elastostatic, or electrical measurements. Lately, hybrid methods that perform different kind of measurements or use different technologies are under development. For example, thermally induced acoustic waves is a new technique for characterizing aerospace materials and structures from remote sensing. To successfully develop this technique for analyzing thermally generated images, new mathematical models to describe the heat flow and the acoustic wave propagation induced by laser or other thermal sources are needed.

Regardless of the kind of measurements taken on the boundary, many different inversion approaches are used in nondestructive testing of industrial components in order to locate any cavities, inclusions, or discontinuities. There are two classes of methods: non-iterative methods and iterative methods. Non-iterative methods, as the reciprocity gap technique for straight-line cracks or the factorization method are fast but do not provide detail information about the crack. On the other hand, iterative methods are computationally more expensive. They need to solve a direct problem at each step to compute the Fréchet derivatives of the direct operator but they usually offer very accurate results (if an initial good estimate of the crack is used). Neural network operators can also be used for defect detection.

The chapter by Ana Carpio and Maria Luisa Rapún considers the inverse scattering problem for the detection of objects buried in a medium where waves are described by the Helmholtz equation. The task is to find the locations and shapes of the targets, given many input and output waves. Extensions to the problem in which the objects are buried in a elastic medium is also given.

History Matching

Reservoir simulators are routinely used to predict and optimize oil production. The accuracy of the numerical solution computed by the simulation depends on the accuracy of the physical parameters that characterize the reservoir and, unfortunately, these are poorly known due to the relative inaccessibility of the reservoir to sampling. Therefore, proper characterization of the petroleum reservoir heterogeneity is a crucial aspect of any optimal reservoir management strategy. It helps to better understand the reservoir behavior so that its performance can be predicted and controlled with higher reliability. The history matching problem consists in adjusting a set of parameters, such as the permeability and porosity distributions, in order to match the data obtained with the simulator to the actual production data in the reservoir. The input data used to solve the inverse problem in history matching usually consist of the pressure or the flow at the production wells.

One important difference between the history matching problem and other applications such as X-ray tomography or magnetic resonance imaging is that in the history matching problem the experimental setup cannot easily be changed in order to obtain independent data. Typically, only one field experiment is available due to the production process. A small reservoir model might have 50,000 to 100,000 grid blocks and solutions to the history matching problem are far from being unique. As a consequence, production data rarely suffices to characterize reservoir heterogeneity and it is, therefore, desirable to integrate all other available data into the model, such as geological or seismic interpretations, to reduce the amount of uncertainty. Lately, several researchers are making a lot of effort to incorporate geological data, coming from sparse measurements of the permeability, as constrains to the inverse problem. Most of the approaches are formulated to honor the histograms generated from these data.

3 Stability

Let us consider the Hilbert spaces P and D (the parameter and data spaces, respectively) and the linear bounded operator $A : P \rightarrow D$. The goal of any imaging technique is to take a set of data $d \in D$ and produce an image of some (unknown) physical quantity of interest $f \in P$. Mathematically, this is done by solving an inverse problem represented in abstract form by

$$A f = d. \tag{2}$$

However, the inverse problems arising from image reconstruction are usually ill-posed in the sense of Hadamard. A problem is well-posed in the sense of Hadamard if:

1. A solution exists.
2. The solution is unique.
3. The solution depends continuously on the given data.

Otherwise, the problem is ill-posed. A good measure of the degree of ill-posedness is given by the singular value decomposition (SVD) of A. The faster the decay rate of the singular values, the worse the degree of ill-posedness. If the singular values decay asymptotically at an exponential rate, the inverse problem is severely ill-posed. This is the case, for example, of diffuse optical tomography or cross-borehole electromagnetic induction tomography.

The most important issue in image reconstruction, from the computational point of view, is the failure of the third property. One can deal with the first two by approximate inverses such as the well known Moore–Penrose pseudoinverse $A^{\dagger}$, but if the third property is not fulfilled any *simple* image reconstruction algorithm will lead to instabilities and the image will not be reliable. To impose stability on an ill-posed inverse problem we must introduce continuity in the inverse operator A^{-1} by means of a regularization method [4].

We stress here that any measurement procedure leads to data perturbed by noise. Rather than attempting to invert A directly (perfect inverses in realistic problems, especially those arising in imaging reconstruction, typically never happen), we adopt a least square approach and seek a solution f that minimizes the cost functional

$$J(f) = \frac{1}{2} \|R(f)\|_2^2, \tag{3}$$

where $R(f) = A f - d$ is the residual operator which measures the data misfit.

The basic idea of regularization is to replace the original ill-posed problem by a nearby well-posed problem. There are two standard methods to accomplish regularization. The first method, Tikhonov–Phillips regularization, involves adding the identity operator (or other simple operator), multiplied by a small factor called the regularization parameter, to the approximate inverse of A. The choice of an optimal, or at least good, regularization parameter is not trivial at all, and depends on the specific imaging technique we consider. While a regularization parameter too small do not restore the continuous dependence of the solution upon the data, a too large one compromises the resolution which might lead to a useless image for practical purposes. The second method, truncated SVD, discard the smallest singular vectors corresponding to singular values below some small threshold ϵ. This eliminates the impact of the noise-like data on the final image. The regularization parameter here is ϵ.

One way to introduce regularization is to change the cost functional (3) by incorporating an additional term. From this point of view, Tikhonov–Phillips regularization amounts to minimize the cost functional

$$J(f) = \frac{1}{2}\|R(f)\|_2^2 + \lambda\|Q(f)\|_2^2 \tag{4}$$

where $\lambda > 0$ is the regularization parameter. The second term in (4) is the penalty term through which prior information about the unknown f can be taken into account. The prior information operators most often used in image reconstruction are the first and second differential operators. The corresponding implicit prior assumptions made when using these operators are that f changes slowly or f is smooth, respectively.

Tikhonov–Phillips type regularizations, therefore, changes the cost functional significantly which may not be desirable in some applications. In fact, this might compromise the resolution of the final image. By adding a operator proportional to the identity to the approximate inverse, Tikhonov–Phillips regularization effectively shifts all the singular values away from zero, thereby reducing the influence of the smallest singular values (noise-like) on the final reconstructed image but, at the same time, reducing the resolvability of the largest ones (signal-like). On the other hand, truncated SVD regularization just set the smallest singular values to zero, so they are preferred when resolution is the only issue. We note, however, that SVD is computationally highly demanding, and therefore, they are not applied for large scale problems (as it is often the case in realistic inverse problems).

The continue quest for methods that can improve resolution beyond the *natural* or *expected* resolution limits is a major issue in imaging. Another possible approach to regularization is to work with the original cost functional (3) and introduce constraints on f requiring that the solution belongs to a smaller subset of the parameter space P. This alternative form of regularization is presented in the chapter by Oliver Dorn, Hugo Bertete-Aguirre, and George Papanicolaou. Two different and interesting interpretations of this regularization approach are also given.

In this same spirit, a similar strategy is to incorporate parameter space constraints so that the image exhibit the desirable properties. Lately, there is a lot of interest in shape-reconstruction techniques that provide well-defined boundaries of the objects to be imaged. Besides, these techniques incorporate an additional intrinsic regularization (by constraining the parameter space due to prior knowledge) that reduces significantly the dimensionality of the inverse problem, and thereby stabilizing the reconstruction.

4 Image Reconstruction Techniques

Although the different imaging modalities presented in the previous section probe different parameters or characteristics that describe the media, and work under different physical principles, they all share the same mathematical foundations, and the tools used to solve the associated inverse problem have a great deal in common. Here, we will first consider two cases in which the applied imaging techniques can be seen as different. In the last subsection we briefly mention other problems and image reconstruction techniques that the reader can find in the lectures of this book.

4.1 Computerized Tomography

Image reconstruction from projections is a linear inverse problem (typically underdetermined) that attempts to retrieve an unknown function (the image) from a set of lower dimensional profiles of this unknown function. In several clinical imaging techniques such as computerized tomography, magnetic resonance imaging, positron emission tomography, or single particle emission computed tomography, these profiles are obtained by integrating the unknown function along straight lines orthogonal to the projections. As a consequence, they all share similar imaging algorithms. Some of this techniques, as the backprojection algorithm, the Kaczmarz's method, or Fourier methods are outlined in the lecture by Frank Natterer in this book.

Mathematically, the inverse problem can be formulated as follows. Let us assume that the set of measurements

$$(Rf)(\theta, s) = \int_{\mathbb{R}^2} f(x, y) \delta(x \cos \theta + y \sin \theta - s) \, dx \, dy \qquad (5)$$

over $s \in \mathbb{R}$ is available for some projections $\theta \in [0, \pi]$, where δ is the Dirac's delta function. The inverse problem is to find f from a finite number of measurements $(Rf)(\theta_i, s_j) = d_{ij}$. If Rf is available for all θ and s (without noise), then the inverse Radon transform solves the problem exactly. This inverse problem is, nevertheless, ill posed.

Simple backprojection is conceptually quite easy but it does not correctly solve this inverse problem. Simple backprojection reconstructs an image by taking each projection and smearing (or backprojecting) it along the line it was acquired. As a result of this simple procedure, the backprojected image is blurred. Filtered backprojection, which is the most important reconstruction algorithm in tomography, is a technique devised to correct the blurring encountered in simple backprojection. Each projection is properly filtered before the backprojection is done to counteract the blur. The image generated by this algorithm agrees with the *true* image if an infinite number of projections are available. The filtered backprojection can actually be viewed as a numerical implementation of the inverse Radon transform which solves the inverse problem exactly.

The Kaczmarz's method, also called algebraic reconstruction technique (ART), is an iterative technique. Iterative techniques may be slow, but they are generally used when better algorithms are not available. In fact, they were used in the first commercial medical CT scanner in 1972. The ART algorithm uses successive orthogonal projections to approximate solutions to systems of linear equations. This algorithm approximates a solution by projecting a given vector successively onto the hyperplanes defined by the individual equations. To start the ART algorithm, all the pixels in the image are set to an initial arbitrary value.

Finally, Fourier methods takes the two-dimensional Fourier transform of the image and the one-dimensional Fourier transform of each of its views. It turns out that the relationship between an image and its views is much simpler in the frequency domain than in the spatial domain.

4.2 Diffuse Optical Tomography

Diffuse optical tomography is an example of an imaging modality that requires the solution of a nonlinear inverse problem. This kind of imaging techniques are still the subject of a lot of academic research and their future usefulness in areas such as medicine are still under discussion.

Several challenges arise in diffuse optical tomography due to the multiple scattering of light in biological tissue. Physically, multiple scattering causes severe image blurring. Hence, one can not make use of direct images. Since light does not travel along straight lines, one can not make use of the reconstruction algorithms applied in X-ray tomography neither. Rather, one must develop methods to reconstruct images from strongly scattered light measurements. One possible approach is to discriminate between weakly and strongly scattered photons. Various techniques, such as gating techniques, have been devised to carry out this task. Time gating uses picosecond time-resolved techniques to select the early arriving photons at the detector, thereby selecting those photons traveling along straight paths. Therefore, the information contained in the detected signal give better contrast and/or could be used in X-ray tomography based imaging algorithms. Polarization gating is based on the fact that weakly scattered photons (carrying the useful information) retains the polarization state of the incident light while strongly scattered photons do not. In this book, the reader can find a study by Miguel Moscoso that uses polarization effects to obtain enhanced resolution images in biological tissue.

4.3 Other Problems

Image reconstruction methods using a backpropagation strategy based on adjoint fields have been applied in diffuse optical tomography. The reader can find an exposition of this method for the cross-borehole electromagnetic induction tomography application in the lecture by Oliver Dorn, Hugo Bertete-Aguirre, and George Papanicolaou. The adjoint field technique is an iterative method that approximately solves the inverse problem (upon linearization) by making use of the same forward modeling code two time in each step of the algorithm: one to compute the misfit between the 'real' and the simulated data, and one to compute the sensitivity map.

Ana Carpio and María Luisa Rapún explain a promising numerical scheme for solving inverse scattering problems by means of a topological derivative method. The topological derivative is essentially the gradient of an objective functional in the limit of a hole vanishing to zero. Thereby, it provides information about the benefits of introducing a hole at a given location without the need of a further analysis. Lately, there is a lot of interest in the use of topological derivatives for computationally efficient methods for shape functional optimization problems.

An example of coherent imaging can be found in the lecture by Oliver Dorn. He establishes a direct link between the time-reversal technique and the adjoint method for imaging. Using this relationship, he derives new solution strategies for an inverse problem in telecommunication. These strategies are all based on iterative time-reversal experiments, which try to solve the inverse problem *experimentally* instead of computationally. He focuses, in particular, on examples from underwater acoustic communication and wireless communication in a Multiple-Input Multiple-Output (MIMO) setup.

Acknowledgements

The author acknowledges support from the Spanish Ministerio de Educación y Ciencia (grants n° FIS2004-03767 and FIS2007-62673) and from the European Union (grant LSHG-CT-2003-503259).

References

1. Claerbout J.F., *Fundamentals of Geophysical Data Processing*, McGraw-Hill international series in the earth and planetary sciences, 1976.
2. Epstein C.L., *Introduction to the Mathematics of Medical Imaging*, Prentice Hall, 2003.
3. Hellier C., *Handbook of nondestructive evaluation*, McGraw-Hill, 2001.
4. Kirsch A., *An introduction to the mathematical theory of inverse problems*, Applied Mathematical Science vol. 120, Springer, 1996.
5. Natterer F. and Wübbeling F., *Mathematical Methods in Image Reconstruction*, SIAM Monographs on Mathematical Modeling and Computation, 2001.

X-ray Tomography

Frank Natterer

Institut für Numerische und Angewandte Mathematik, University of Münster,
Einsteinstraße 62, 48149 Münster, Germany
natterer@math.uni-muenster.de

Summary. We give a survey on the mathematics of computerized tomography. We start with a short introduction to integral geometry, concentrating on inversion formulas, stability, and ranges. We then go over to inversion algorithms. We give a detailed analysis of the filtered backprojection algorithm in the light of the sampling theorem. We also describe the convergence properties of iterative algorithms. We shortly mention Fourier based algorithms and the recent progresses made in their accurate implementation. We conclude with the basics of algorithms for cone beam scanning which is the standard scanning mode in present days clinical practice.

1 Introduction

X-ray tomography, or computerized tomography (CT) is a technique for imaging cross sections or slices (slice = $\tau o\mu os$ (greek)) of the human body. It was introduced in clinical practice in the 70s of the last century and has revolutionized clinical radiology. See Webb (1990), Natterer and Ritman (2002) for the history of CT.

The mathematical problem behind CT is the reconstruction of a function f in $\mathbb{R}^2$ from the set of its line integrals. This is a special case of integral geometry, i.e. the reconstruction of a function from integrals over lower dimensional manifolds. Thus, integral geometry is the backbone of the mathematical theory of CT.

CT was soon followed by other imaging modalities, such as emission tomography (single particle emission tomography (SPECT) and positron emission tomography (PET)), magnetic resonance imaging (MRI). CT also found applications in various branches of science and technology, e.g. in seismics, radar, electron microscopy, and flow. Standard references are Herman (1980), Kak and Slaney (1987), Natterer and Wübbeling (2001).

In these lectures we give an introduction into the mathematical theory and the reconstruction algorithms of CT. We also discuss matters of resolution and stability. Necessary prerequisites are calculus in $\mathbb{R}^n$, Fourier transforms, and the elements of sampling theory.

Since the advent of CT many imaging techniques have come into being, some of them being only remotely related to the straight line paradigm of CT, such as impedance tomography, optical tomography, and ultrasound transmission tomography. For the study of these advanced technique a thorough understanding of CT is at least useful. Many of the fundamental tools and issues of CT, such as backprojection, sampling, and high frequency analysis, have their counterparts in the more advanced techniques and are most easily in the framework of CT.

The outline of the paper is as follows. We start with a short account of the relevant parts of integral geometry, with the Radon transform as the central tool. Then we discuss in some detail reconstruction algorithms, in particular the necessary discretizations for methods based on inversion formulas and convergence properties of iterative algorithm. Finally we concentrate on the 3D case which is the subject of current research.

2 Integral Geometry

In this section we introduce the relevant integral transforms, derive inversion formulas, and study the ranges. For a thorough treatment see Gelfand, Graev, and Vilenkin (1965), Helgason (1999), Natterer (1986).

2.1 The Radon Transform

Let f be a function in $\mathbb{R}^n$. To avoid purely technical difficulties we assume f to be smooth and of compact support, if not said otherwise.

For $\theta \in S^{n-1}$, $s \in \mathbb{R}^1$ we define

$$(Rf)(\theta, s) = \int_{x \cdot \theta = s} f(x) dx . \tag{1}$$

R is the Radon transform. dx is the restriction of the Lebesgue measure in $\mathbb{R}^n$ to $x \cdot \theta = s$. Other notations are

$$(Rf)(\theta, s) = \int_{\theta^\perp} f(s\theta + y) dy \tag{2}$$

with $\theta^\perp$ the subspace of $\mathbb{R}^n$ orthogonal to θ, and

$$(Rf)(\theta, s) = \int \delta(x \cdot \theta - s) f(x) dx \tag{3}$$

with δ the Dirac δ-function.

The Radon transform is closely related to the Fourier transform

$$\hat{f}(\xi) = (2\pi)^{-n/2} \int_{\mathbb{R}^n} e^{-ix \cdot \xi} f(x) dx .$$

Namely, we have the so-called projection-slice theorem.

Theorem 2.1.
$$(Rf)^\wedge(\theta,\sigma) = (2\pi)^{(n-1)/2}\hat{f}(\sigma\theta) .$$

Note that the Fourier transform on the left hand side is the 1D Fourier transform of Rf with respect to its second variable, while the Fourier transform on the right hand side is the nD Fourier transform of f.

For the proof of Theorem 2.1 we make use of the definition (2) of Radon transform, yielding

$$(Rf)^\wedge(\theta,\sigma) = (2\pi)^{-1/2}\int_{\mathbb{R}^1} e^{-is\sigma}(Rf)(\theta,s)ds$$

$$= (2\pi)^{-1/2}\int_{\mathbb{R}^1} e^{-is\sigma}\int_{\theta^\perp} f(s\theta+y)dyds .$$

Putting $x = s\theta + y$, i.e. $s = x \cdot \theta$, we have

$$(Rf)^\wedge(\theta,s) = (2\pi)^{-1/2}\int_{\mathbb{R}^n} e^{-i\sigma x\cdot\theta}f(x)dx$$

$$= (2\pi)^{(n-1)/2}\hat{f}(\sigma\theta) .$$

The proofs of many of the following results follow a similar pattern. So we omit proofs unless more sophisticated tools are needed.

It is possible to extend R to a bounded operator $R : L_2(\mathbb{R}^n) \to L_2(S^{n-1}\times\mathbb{R}^1)$. As such R has an adjoint $R^* : L_2(S^{n-1} \times \mathbb{R}^1) \to L_2(\mathbb{R}^n)$ which is easily seen to be

$$(R^*g)(x) = \int_{S^{n-1}} g(\theta, x \cdot \theta)d\theta . \tag{4}$$

R^* is called the backprojection operator in the imaging literature.

We also will make use of the Hilbert transform

$$(Hf)(s) = \frac{1}{\pi}\int \frac{f(t)}{s-t}dt \tag{5}$$

which, in Fourier domain, is given by

$$(Hf)^\wedge(\sigma) = -i\, sgn(\sigma)\hat{f}(\sigma) \tag{6}$$

with $sgn(\sigma)$ the sign of the real number σ. Now we are ready to state and to prove Radon's inversion formula:

Theorem 2.2. *Let* $g = Rf$. *Then*

$$f = \frac{1}{2}(2\pi)^{1-n}R^*H^{n-1}g^{(n-1)}$$

where $g^{(n-1)}$ *stands for the derivative of order* $n - 1$ *of* g *with respect to the second argument.*

For the proof we write the Fourier inversion formula in polar coordinates and make use of Theorem 2.1 and the evenness of g, i.e. $g(\theta, s) = g(-\theta, -s)$.

Special cases of Theorem 2.2 are

$$f(x) = \frac{1}{4\pi^2} \int\limits_{S^1} \int\limits_{\mathbb{R}^1} \frac{g'(\theta, s)}{x \cdot \theta - s} ds d\theta \tag{7}$$

for $n = 2$ and

$$f(x) = -\frac{1}{8\pi^2} \int\limits_{S^2} g''(\theta, x \cdot \theta) d\theta \tag{8}$$

for $n = 3$. Note that there is a distinctive difference between (7) and (8): The latter one is local in the sense that for the reconstruction of f at x_0 only the integrals $g(\theta, s)$ over those planes $x \cdot \theta = s$ are needed that pass through x_0 or nearby. In contrast, (7) is not local in this sense, due to the integral over $\mathbb{R}^1$.

In order to study the stability of the inversion process we introduce the Sobolev spaces H^α on $\mathbb{R}^n$, $S^{n-1} \times \mathbb{R}^1$ with norms

$$\|f\|^2_{H^\alpha(\mathbb{R}^n)} = \int\limits_{\mathbb{R}^n} (1 + |\xi|^2)^\alpha |\hat{f}(\xi)|^2 d\xi \,,$$

$$\|g\|^2_{H^\alpha(S^{n-1} \times \mathbb{R}^1)} = \int\limits_{S^{n-1}} \int\limits_{\mathbb{R}^1} (1 + \sigma^2)^\alpha |\hat{g}(\theta, \sigma)|^2 d\sigma d\theta \,.$$

As in Theorem 2.1, the Fourier transform $\hat{g}$ is the 1D Fourier transform of g with respect to the second argument.

Theorem 2.3. *There exist constants $c, C > 0$ depending only on α, n, such that*

$$c\|f\|_{H^\alpha(\mathbb{R}^n)} \leq \|Rf\|_{H^{\alpha+(n-1)/2}(S^{n-1} \times \mathbb{R}^1)} \leq C\|f\|_{H^\alpha(\mathbb{R}^n)}$$

for $f \in H^\alpha(\mathbb{R}^n)$ with support in $|x| < 1$.

The conclusion is that Rf is smoother than f by the order $(n-1)/2$. This indicates that the reconstruction process, i.e. the inversion of R, is slightly ill-posed.

Finally we study the range of R.

Theorem 2.4. *For $m \geq 0$ an integer let*

$$p_m(\theta) = \int\limits_{\mathbb{R}^1} s^m (Rf)(\theta, s) ds \,.$$

Then, p_m is a homogeneous polynomial of degree m in θ.

The proof is by verification. The question wether or not the condition of Theorem 2.4 is sufficient for a function g of θ, s being in the range of R is the subject of the famous Helgason–Ludwig theorem.

Theorem 2.4 has applications to cases in which the data function g is not fully specified.

2.2 The Ray Transform

For f a function in $\mathbb{R}^n$, $\theta \in S^{n-1}$, $x \perp \theta^\perp$ we define

$$(Pf)(\theta, x) = \int_{\mathbb{R}^1} f(x + t\theta)dt \tag{9}$$

P is called the ray (or X-ray) transform. The projection-slice theorem for P reads

Theorem 2.5.

$$(Pf)^\wedge(\theta, \xi) = (2\pi)^{1/2}\hat{f}(\xi), \xi \in \theta^\perp .$$

The Fourier transform on the left hand side is the $(n-1)D$ Fourier transform in $\theta^\perp$, while the Fourier transform on the right hand side is the nD Fourier transform in $\mathbb{R}^n$.

The adjoint ray transform, which we call backprojection again, is given by

$$(P^*g)(x) = \int_{S^{n-1}} g(\theta, E_\theta x)d\theta \tag{10}$$

with E_θ the orthogonal projection onto $\theta^\perp$, i.e. $E_\theta x = x - (x \cdot \theta)\theta$. We have the following analogue of Radon's inversion formula:

Theorem 2.6. *Let $g = Pf$. Then*

$$f = \frac{1}{2\pi|S^{n-1}|} P^* I^{-1}g$$

with the Riesz potential

$$(I^{-1}g)^\wedge(\xi) = |\xi|\hat{g}(\xi)$$

in $\theta^\perp$.

Theorem 2.6 is not as useful as Theorem 2.2. The reason is that for $n \geq 3$ (for $n = 2$ Theorems 2.2 and 2.6 are equivalent) the dimension $2(n-1)$ of the data function g is greater than the dimension n of f. Hence the problem of inverting P is vastly overdetermined. A useful formula would compute f from the values $g(\theta, \cdot)$ where θ is restricted to some set $S_0 \subseteq S^{n-1}$. Under which conditions on S_0 is f uniquely determined?

Theorem 2.7. *Let $n = 3$, and assume that each great circle on S^2 meets S_0. Then, f is uniquely determined by $g(\theta, \cdot)$, $\theta \in S_0$.*

The condition on S_0 in Theorem 2.7 is called Orlov's completeness condition. In the light of Theorem 2.5, the proof of Theorem 2.7 is almost trivial: Let $\xi \in \mathbb{R}^3 \setminus \{0\}$. According to Orlov's condition there exists $\theta \in S_0$ on the great circle $\xi^\perp \cap S^2$. For this θ, $\xi \in \theta^\perp$, hence

$$\hat{f}(\xi) = (2\pi)^{-1/2}\hat{g}(\theta, \xi)$$

by Theorem 2.5. Thus $\hat{f}$ is uniquely (and stably) determined by g on S_0.

An obvious example for a set $S_0 \subseteq S^2$ satisfying Orlov's completeness condition is a great circle. In that case inversion of P is simply 2D Radon inversion on planes parallel to S_0. Orlov gave an inversion formula for S_0 a spherical zone.

2.3 The Cone-Beam Transform

The relevant integral transform for fully 3D X-ray tomography is the cone beam transform in $\mathbb{R}^3$

$$(Cf)(a,\theta) = \int_0^\infty f(a + t\theta)dt \tag{11}$$

where $\theta \in S^2$. The main difference to the ray transform P is that the source point a is restricted to a source curve A surrounding the object to be imaged.

Of course the following question arises: Which condition on A guarantees that f is uniquely determined by $(Cf)(a, \cdot)$?

A first answer is: Whenever A has an accumulation point outside supp (f) (provided f is sufficiently regular). Unfortunately, inversion under this assumptions is based on analytic continuation, a process that is notoriously unstable. So this result is useless for practical reconstruction.

In order to find a stable reconstruction method (for suitable source curves A) we make use of a relationship between C and R discovered by Grangeat (1991):

Theorem 2.8.

$$\frac{\partial}{\partial s}(Rf)(\theta, s)\,|_{s=a\cdot\theta} = \int_{\theta^\perp \cap S^2} \frac{\partial}{\partial\theta}(Cf)(a,\omega)d\omega \ .$$

The notation on the right hand side needs explication: $(Cf)(a, \cdot)$ is a function on S^2. θ is a tangent vector to S^2 for each $\omega \in S^2 \cap \theta^\perp$. Thus the derivative $\frac{\partial}{\partial\theta}$ makes sense on $S^2 \cap \theta^\perp$.

Since Theorem 2.8 is the basis for present day's work on 3D reconstruction we sketch two proofs. The first one is completely elementary. Obviously it suffices to put $\theta = e_3$, the third unit vector, in which case it reads

$$\frac{\partial}{\partial s}(Rf)(\theta, a_3) = \int_{\omega \in S^1} \left[\frac{\partial}{\partial z}(Cf)\left(a, \begin{pmatrix} \omega \\ z \end{pmatrix} \right) \right]_{z=0} d\omega$$

where a_3 is the third component of the source point a. It is easy to verify that right- and left-hand side both coincide with

$$\int_{\mathbb{R}^2} \frac{\partial f}{\partial x_3} \left(a + \begin{pmatrix} x' \\ 0 \end{pmatrix} \right) dx' \ .$$

The second proof makes use of the formula

$$\int_{S^{n-1}} (Cf)(a,\omega)h(\theta\cdot\omega)d\omega = \int_{\mathbb{R}^1} (Rf)(\theta,s)h(s-a\cdot\theta)ds$$

that holds – for suitable functions f in $\mathbb{R}^2$ – for h a function in $\mathbb{R}^1$ homogeneous of degree $1-n$; see Hamaker et al. (1980).

Putting $h=\delta'$ yields Theorem 2.8.

Theorem 2.9. *Let the source curve A satisfy the following condition: Each plane meeting supp (f) intersects A transversally. Then, f is uniquely (and stably) determined by $(Cf)(a,\theta)$, $a\in A$, $\theta\in S^2$.*

The condition on A in Theorem 2.9 is called the Kirillov–Tuy condition.

The proof of Theorem 2.9 starts out from Radon's inversion formula (8) for $n=3$:

$$f(x) = -\frac{1}{8\pi^2}\int_{S^2}\left[\frac{\partial^2}{\partial s^2}Rf(\theta,s)\right]_{s=x\cdot\theta}d\theta \tag{12}$$

Now assume A satisfies the Kirillov–Tuy condition, and let $x\cdot\theta = s$ be a plane hitting supp (f). Then there exists $a\in A$ such that $a\cdot\theta = s$, and

$$\left[\frac{\partial}{\partial s}Rf(\theta,s)\right]_{s=x\cdot\theta} = \left[\frac{\partial}{\partial s}Rf(\theta,s)\right]_{s=a\cdot\theta}$$

$$= \int_{\theta^\perp\cap S^2}\frac{\partial}{\partial\theta}(Cf)(a,\omega)d\omega$$

is known by Theorem 2.8. Since A intersects the plane $x\cdot\theta = s$ transversally this applies also to the second derivative in (12). Hence the integral in (12) can be computed from the known values of Cf for all planes $x\cdot\theta = s$ hitting supp (f), and for the others it is zero.

Theorem 2.8 in conjunction with Radon's inversion formula (8) is the basis for reconstruction formulas for C with source curves A satisfying the Kirillov–Tuy condition. The simplest source curve one can think of, a circle around the object, does not satisfy the Kirillov–Tuy condition. For a circular source curve an approximate inversion formula, the FDK formula, exists; see Feldkamp et al. (1984). In medical applications the source curve is a helix.

2.4 The Attenuated Radon Transform

Let $n=2$. The attenuated Radon transform of f is defined to be

$$(R_\mu f)(\theta,s) = \int_{x\cdot\theta=s} e^{-(C_\mu)(x,\theta_\perp)}f(x)dx$$

where μ is another function in $\mathbb{R}^2$ and $\theta_\perp$ is the unit vector perpendicular to θ such that det $(\theta, \theta_\perp) = 1$. This transform comes up in SPECT, where f the (sought-for) activity distribution and μ the attenuation map.

R_μ admits an inversion formula very similar to Radon's inversion formula for R:

Theorem 2.10. *Let $g = R_\mu f$. Then,*

$$f = \frac{1}{4\pi} Re\ div\ R^*_{-\mu}(\theta e^{-h} H e^h g)$$

where

$$h = \frac{1}{2}(I + iH)R\mu\ ,$$

$$(R^*_\mu g)(x) = \int_{S^1} e^{-(C\mu)(x,\theta_\perp)} g(\theta, x \cdot \theta) d\theta\ .$$

This is Novikov's inversion formula; see Novikov (2000). Note that the back-projection $R^*_{-\mu}$ is applied to a vector valued function. For $\mu = 0$ Novikov's formula reduces to Radon's inversion formula (7).

There is also an extension of Theorem 2.4 to the attenuated case:

Theorem 2.11. *Let $k > m \geq 0$ integer and h the function from Theorem 2.10. Then,*

$$\int_{\mathbb{R}^1} \int_{S^1} s^m e^{ik\varphi + h(\theta,s)} (R_\mu f)(\theta, s) d\theta ds = 0$$

where $\theta = \begin{pmatrix} \cos\varphi \\ \sin\varphi \end{pmatrix}$.

The proofs of Theorems 2.10 and 2.11 are quite involved. They are both based on the following remark: Let

$$u(x, \theta) = h(\theta, x \cdot \theta) - (C\mu)(x, \theta_\perp)\ .$$

Then, u admits a Fourier series representation as

$$u(x, \theta) = \sum_{\ell > 0 \text{ odd}} u_\ell(x) e^{i\ell\varphi}$$

with certain functions $u_\ell(x)$ and $\theta = \begin{pmatrix} \cos\varphi \\ \sin\varphi \end{pmatrix}$.

Of course this amounts to saying that $u(x, \cdot)$ admits an analytic continuation from S^1 into the unit disk.

2.5 Vectorial Transforms

Now let f be a vector field. The scalar transforms P, R can be applied componentwise, so Pf, Rf make sense. They give rise to the vectorial ray transform

$$(\mathcal{P}f)(\theta,s) = \theta \cdot (Pf)(\theta,s) \, , \quad x \in \theta^{\perp}$$

and the Radon normal transform

$$(\mathcal{R}^{\perp}f)(\theta,s) = \theta \cdot (Rf)(\theta,s) \, .$$

Obviously, vectorial transforms can't be invertible. In fact their null spaces are huge.

Theorem 2.12. $\mathcal{P}f = 0$ *if and only if* $f = \nabla\psi$ *for some* ψ.

The nontrivial part of the proof makes use of Theorem 2.5: If $\mathcal{P}f = 0$, then, for $\xi \in \theta^{\perp}$

$$(\mathcal{P}f)^{\wedge}(\theta,\xi) = \theta \cdot (Pf)^{\wedge}(\theta,\xi) = (2\pi)^{1/2}\theta \cdot \hat{f}(\xi) = 0 \, .$$

Hence $\theta^{\perp}$ is invariant under $\hat{f}$. This is only possible if $\hat{f}(\xi) = i\hat{\psi}(\xi)\xi$ with some $\hat{\psi}(\xi)$. It remains to show that $\hat{\psi}$ is sufficiently regular to conclude that $f = \nabla\psi$.
A similar proof leads to

Theorem 2.13. *Let* $n = 3$. $\mathcal{R}^{\perp}f = 0$ *if and only if* $f = \text{curl } \phi$ *for some* ϕ.

Inversion formulas can be derived for those parts of the solution that are uniquely determined. We make use of the Helmholtz decomposition to write a vector f in the form

$$f = f^s + f^i$$

where f^s is the solenoidal part (i.e. div $f^s = 0$ or $f^s = \text{curl } \phi$) and f^i is the irrotational part (i.e. curl $f^i = 0$, $f^i = \nabla\psi$). Then, Theorem 2.12 states that the solenoidal part of f is uniquely determined by $\mathcal{P}f$, while the irrotational part of f is uniquely determined by $\mathcal{R}^{\perp}f$.
We have the following inversion formula:

Theorem 2.14. *Let* $g = \mathcal{P}f$ *and let* f *be solenoidal (i.e.* $f^i = 0$). *Then,*

$$f = \frac{n-1}{2\pi|S^{n-2}|} \mathcal{P}^* I^{-1}g \, ,$$

$$(\mathcal{P}^*g)(x) = \int_{S^{n-1}} g(\theta, E_\theta x)\theta d\theta$$

and I^{-1} *is the Riesz potential already used in Theorem 2.6.*

For more results see Sharafutdinov (1994).

3 Reconstruction Algorithms

There are essentially three classes of reconstruction algorithms. The first class is based on exact or approximate inversion formulas, such as Radon's inversion formula. The prime example is the filtered backprojection algorithm. It's not only the work horse in clinical radiology, but also the model for many algorithms in other fields. The second class consists of iterative methods, with Kaczmarz's method (called algebraic reconstruction technique (ART) in tomography) as prime example. The third class is usually called Fourier methods, even though Fourier techniques are used also in the first class. Fourier methods are direct implementations of the projection slice theorem (Theorems 2.1, 2.5) without any reference to integral geometry.

3.1 The Filtered Backprojection Algorithm

This can be viewed as an implementation of Radon's inversion formula (Theorem 2.2). However it is more convenient to start out from the formula

Theorem 3.1. *Let $V = R^*v$. Then,*

$$V * f = R^*(v * g)$$

where the convolution on the left hand side is in $\mathbb{R}^n$, i.e.

$$(V * f)(x) = \int_{\mathbb{R}^n} V(x - y)f(y)dy$$

and the convolution on the right hand side is in $\mathbb{R}^1$, i.e.

$$(v * g)(\theta, s) = \int_{\mathbb{R}^1} v(\theta, s - t)g(\theta, t)dt \ .$$

The proof of Theorem 3.1 require not much more than the definition of R^*.

Theorem 3.1 is turned into an (approximate) inversion formula for R by choosing $V \sim \delta$. Obviously V is the result of the reconstruction procedure for $f = \delta$. Hence V is the point spread function, i.e. the function the algorithm would reconstruct if the true f were the δ-function.

Theorem 3.2. *Let $V = R^*v$ with v independent of θ. Then,*

$$\hat{V}(\xi) = 2(2\pi)^{(n-1)/2}|\xi|^{1-n}\hat{v}(|\xi|) \ .$$

For V to be close to δ we need $\hat{V}$ to be close to $(2\pi)^{-\frac{n}{2}}$, hence

$$\hat{v}(\sigma) \sim \frac{1}{2}(2\pi)^{-n+1/2}\sigma^{n-1} \ .$$

Let ϕ be a low pass filter, i.e. $\phi(\sigma) = 0$ for $|\sigma| > 1$ and, for convenience, ϕ even. Let Ω be the (spatial) bandwidth. By Shannon's sampling theorem Ω corresponds to the (spatial) resolution $2\pi/\Omega$:

Theorem 3.3. *Let f be Ω-bandlimited, i.e. $\hat{f}(\xi) = 0$ for $|\xi| > \Omega$. Then, f is uniquely determined by $f(h\ell)$, $\ell \in \mathbb{Z}^n$, $h \leq \pi/\Omega$. Moreover, if $h \leq 2\pi/\Omega$, then*

$$\int_{\mathbb{R}^n} f(x)dx = h^n \sum_{\ell} f(h\ell) .$$

An example for a Ω-bandlimited function in $\mathbb{R}^1$ is $f(x) = \text{sinc}(\Omega x)$ where $\text{sinc}(x) = \sin(x)/x$ is the so-called sinc-function. From looking at the graph of f we conclude that bandwidth Ω corresponds to resolution $2\pi/\Omega$.

We remark that Theorem 3.3 has the following corollary: Assume that f, g are Ω-bandlimited. Then,

$$\int_{\mathbb{R}^n} f(x)g(x)dx = h^n \sum_{\ell} f(h\ell)g(h\ell)$$

provided that $h \leq \pi/\Omega$. This is true since fg has bandwidth 2Ω.

We put

$$\hat{v}(\sigma) = 2(2\pi)^{(n-1)/2}|\sigma|^{n-1}\phi(\sigma/\Omega) .$$

v is called the reconstruction filter. As an example we put $n = 2$ and take as ϕ the ideal low pass, i.e.

$$\phi(\sigma) = \begin{cases} 1 , & |\sigma| \leq 1 , \\ 0 , & |\sigma| > 1 . \end{cases}$$

Then,

$$v(s) = \frac{\Omega^2}{4\pi^2}u(\Omega s) , \quad u(s) = \text{sinc}(s) - \frac{1}{2}\left(\text{sinc}\left(\frac{s}{2}\right)\right)^2 .$$

Now we describe the filtered backprojection algorithm for the reconstruction of the function f from $g = Rf$. We assume that f has the following properties:

$$f \text{ is supported in } |x| \leq \rho \tag{13}$$

$$f \text{ is essentially } \Omega\text{-bandlimited, i.e.} \tag{14}$$

$$\hat{f}(\xi) \text{ is negligible in some sense for } |\xi| > \Omega .$$

Since f is essentially Ω-bandlimited we conclude from Theorem 2.1 that the same is true for g. Hence, by a loose application of the sampling theorem,

$$(v * g)(\theta, s_\ell) = \Delta s \sum_{k} v(\theta, s_\ell - s_k)g(\theta, s_k) \tag{15}$$

where $s_\ell = \ell\Delta s$ and $\Delta s \leq \pi/\Omega$. Equation (15) is the filtering step in the filtered backprojection algorithm. In the backprojection step we compute $R^*(v * g)$ in a discrete setting. We restrict ourselves to the case $n = 2$, i.e.

$$(R^*(v * g))(x) = \int_{S^1} (v * g)(\theta, x \cdot) d\theta .$$

In order to discretize this integral properly we make use of Theorem 3.3 again. For this we have to determine the bandwidth of the function $\varphi \mapsto (v * g)(\theta, x \cdot \theta)$ where $\theta = \begin{pmatrix} \cos \varphi \\ \sin \varphi \end{pmatrix}$.

Theorem 3.4. *For k an integer we have*

$$\int_0^{2\pi} e^{-ik\varphi}(v * g)(\theta, x \cdot \theta) d\varphi = \frac{i^k}{2\pi} \int_{|y|<\rho} f(y) e^{-ik\psi} \int_{-\Omega}^{\Omega} \hat{v}_k(\sigma) J_k(-\sigma|x - y|) d\sigma dy$$

where $\psi = \arg(y)$ and J_k is the Bessel function of order k of the first kind.

All that is needed for the proof is the integral representation

$$J_k(s) = \frac{i^{-k}}{2\pi} \int_0^{2\pi} e^{-ik\varphi + s \cos \varphi} d\varphi$$

of the Bessel function.

By Debye's asymptotic relation (see Abramowitz and Stegun (1970)), $J_k(s)$ is negligibly small for $|s| < |k|$ and $|k|$ large. Hence the left hand side in Theorem 3.4 is negligible for $2\Omega\rho < |k|$. In other words, the function $\varphi \mapsto (v * g)(\theta, x \cdot \theta)$ has essential bandwidth $2\Omega\rho$. According to Theorem 3.3,

$$(R^*(v * g))(x) = \Delta\varphi \sum_j (v * g)(\theta_j, x \cdot \theta_j) \tag{16}$$

where $\theta_j = \begin{pmatrix} \cos \varphi_j \\ \sin \varphi_j \end{pmatrix}$, $\varphi_j = j\Delta\varphi$, provided that $\Delta\varphi \le \pi/\Omega\rho$.

The formulas (16), (17) describe the filtered backprojection algorithm – except for an interpolation step in the second argument. It is generally acknowledged that linear interpolation suffices.

3.2 Iterative Algorithms

The equations one has to solve in CT are usually of the form

$$R_j f = g_j , \qquad j = 1, \ldots, p \tag{17}$$

where R_j are certain linear operators from a Hilbert space into an Hilbert space H_j. For instance, for the 2D problem we have

$$(R_j f)(s) = (Rf)(\theta_j, s) \, . \tag{18}$$

Kaczmarz's method solves linear systems of the kind (18) by successively projecting orthogonally onto the affine subspaces $R_j f = g_j$ of H:

$$f_j = P_{j\,\mathrm{mod}\,p} f_{j-1} \, , \qquad j = 1, 2, \ldots \, .$$

Here, P_j is the orthogonal projection onto $R_j f = g_j$, i.e.

$$P_j f = f - R_j^* (R_j R_j^*)^{-1} (R_j f - g_j) \, .$$

For the application to tomography we replace the operator $R_j R_j^*$ by a simpler one, C_j, and introduce a relaxation parameter ω. Putting

$$\begin{aligned} P_j f &= f - R_j^* C_j^{-1} (R_j f - g_j) \, , \\ P_j^\omega f &= (1 - \omega) f + \omega P_j f \, . \end{aligned}$$

we have the following convergence result.

Theorem 3.5. *Let (17) be consistent. Assume that $C_j \leq R_j R_j^*$ (i.e. C_j hermitian and $C_j - R_j R_j^*$ positive semidefinite) and $0 < \omega < 2$. Then, $f_j = P_{j\,\mathrm{mod}\,p}^\omega f_{j-1}$ converges for $f_0 = 0$ to the solution of (18) with minimal norm.*

The condition $0 < \omega < 2$ is reminiscent of the SOR (successive overrelaxation) method of numerical analysis. In fact Kaczmarz's method can be viewed as an SOR method applied to the linear system $R^* R f = R^* g$ where $R = R_1 \times \cdots \times R_p$.

Theorem 3.5 can be applied directly to tomographic problems. The convergence behaviour depends critically on the choice of ω and of the ordering of (17). We analyse this for the standard case (18). For convenience we assume that f is supported in $|x| < 1$ and $g_j = 0$, $j = 1, \ldots, p$.

The following result is due to Hamaker and Solmon (1978):

Theorem 3.6. *Let $\mathcal{C}_m$ be the linear subspace of $L_2(|x| < 1)$ spanned by $T_m(x \cdot \theta_1), \ldots, T_m(x \cdot \theta_p)$, T_m the first kind Chebysheff polynomials. Then,*

$$P_j \mathcal{C}_m \subseteq \mathcal{C}_m \, .$$

This means that $\mathcal{C}_m$ is an invariant subspace under the iteration of Theorem 3.6. Thus we can study the convergence behaviour on each of these finite dimensional subspaces separately. Let $\rho_m(\omega)$ be the spectral radius of

$$P^\omega = P_1^\omega \cdots P_p^\omega \, .$$

$\rho_m(\omega)$ can be computed numerically:

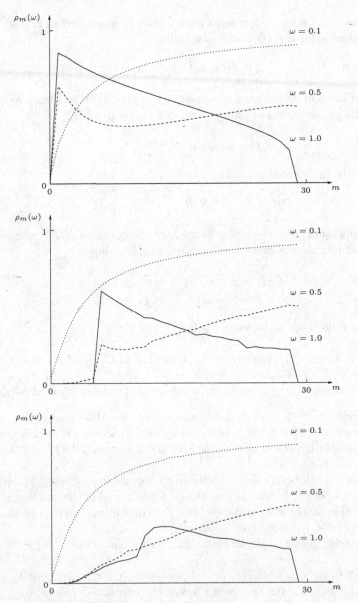

First we consider the solid lines. They correspond to $\omega = 1$. For the sequential ordering $\varphi_j = j\pi/p$, $j = 1, \ldots, p$ the values of $\rho_m(\omega)$ are displayed in the top figure. We see that $\rho_m(\omega)$ is fairly large for low order subspaces $\mathcal{C}_m$, indicating slow convergence for low frequency parts of f, i.e. for the overall features of f. The other figures shows the corresponding values of $\rho_m(\omega)$ for a non-consecutive ordering of the φ_j suggested by Herman and Meyer (1993) and for a random ordering respectively. Both orderings yield much smaller values of $\rho_m(\omega)$, in particular on the low order subspaces. The conclusion is that for practical work one should never use the consecutive ordering, and random ordering is usually as good as any other more sophisticated ordering.

In particular in emission tomography the use of multiplicative iterative methods is quite common. The most popular of these multiplicative algorithms is the expectation-maximization (EM) algorithm: Let $Af = g$ be a linear system, with all the elements of the (not necessarily square) matrix A being non-negative. Assume that A is normalized such that $A1 = 1$ where 1 stands for vectors of suitable dimension containing only 1's. The EM algorithm for the solution of $Af = g$ reads

$$f^{j+1} = f^j A^* \frac{g}{Af^j} , \quad j = 1, 2, \dots ,$$

where division and multiplication are understood componentwise. One can show that f^j converges to the minimizer of the log likelihood function

$$\ell(f) = \sum_i \left(g_i \log(Af)_i - (Af)_i \right) .$$

There exists also a version of the EM algorithm that decomposes the system $Af = g$ into smaller ones, $A_j f = g_j$, $j = 1, \dots, p$, as in the Kaczmarz method (ordered subset EM, OSEM; see Hudson and Larkin (1992)). As in the Kaczmarz method one can obtain good convergence properties by a good choice of the ordering, and the convergence analysis is very much the same.

For more on iterative methods in CT see Censor (1981).

3.3 Fourier Methods

Fourier methods make direct use of relations such as Theorem 2.1. Let's consider the 2D case in which we have

$$\hat{f}(\sigma\theta) = (2\pi)^{-1/2} \hat{g}(\theta, \sigma) , \quad g = Rf . \tag{19}$$

In order to implement this we first have to do a 1D discrete Fourier transform on g for each direction θ_j:

$$\hat{g}(\theta_j, \sigma_\ell) = (2\pi)^{-1/2} \Delta s \sum_k e^{-i\sigma_\ell s_k} g(\theta_j, s_k)$$

where $s_k = k\Delta s$ and $\sigma_\ell = \ell\Delta\sigma$. This provides $\hat{f}$ at the points $\sigma_\ell\theta_j$ which form a grid in polar coordinates. The problem is now to do a 2D inverse Fourier transform on this polar coordinate grid. Various methods have been developed for doing this, in particular the gridding method and various fast Fourier transform on non-equispaced grids; see Fourmont (1999), Beylkin (1995), Steidl (1998).

4 Reconstruction from Cone Beam Data

One of the most urgent needs of present day's X-ray CT is the development of accurate and efficient algorithms for the reconstruction of f from $(Cf)(a, \theta)$ where a is on a source curve A (typically a helix) and $\theta \in S^2$. In practice, θ is restricted to a subset of S_0, but we ignore this additional difficulty.

We assume that the source curve A satisfies Tuy's condition, i.e. A intersects each plane hitting supp (f) transversally. Let A be the curve $x = a(\lambda)$, $\lambda \in \Lambda$. Tuy's condition implies that for each s with $(Rf)(\theta, s) \neq 0$ there is λ such that $s = a(\lambda) \cdot \theta$. In fact there may be more than one such λ, but we assume that their number is finite. Then we may choose a function $M(\theta, s)$ such that

$$\sum_{\substack{\lambda \\ s = a(\lambda) \cdot \theta}} M(\theta, \lambda) = 1$$

for each s. Let $g(\lambda, \theta) = (Cf)(a(\lambda), \theta)$ be the data function. Put

$$G(\lambda, \theta) = \int_{\theta^\perp \cap S^2} \frac{\partial}{\partial \theta} g(\lambda, \omega) d\omega .$$

Then we have

Theorem 4.1. *Let w be a function in $\mathbb{R}^1$ and $V = R^* w'$. Then,*

$$(V * f)(x) = \int_{S^2} \int_\Lambda w((x - a(\lambda)) \cdot \theta) G(\lambda, \theta) |a'(\lambda) \cdot \theta| M(\theta, \lambda) d\lambda d\theta .$$

For the proof we start out from the 3D case of Theorem 3.1:

$$(V * f)(x) = \int_{S^2} \int_{\mathbb{R}^1} w'(x \cdot \theta - s)(Rf)(\theta, s) ds d\theta$$

$$= \int_{S^2} \int_{\mathbb{R}^1} w(x \cdot \theta - s)(Rf)'(\theta, s) ds d\theta .$$

In the s integral we make the substitution $s = a(\lambda) \cdot \theta$, obtaining

$$(V * f)(x) = \int_{S^2} \int_\Lambda w((x - a) \cdot \theta)(Rf)'(\theta, a(\lambda) \cdot \theta) |a'(\lambda) \cdot \theta| M(\theta, \lambda) d\lambda d\theta ,$$

the factor $M(\theta, s)$ being due to the fact that $\lambda \to a(\lambda) \cdot \theta$ is not one-to-one. By Theorem 2.8,

$$(Rf)'(\theta, a(\lambda) \cdot \theta) = G(\lambda, \theta) ,$$

and this finishes the proof.

Theorem 4.1 is the starting point for a filtered backprojection algorithm. For

$$w' = -\frac{1}{8\pi^2}\delta' ,$$

δ the 1D δ-function, we have $V = \delta$, the 3D δ-function, hence

$$f(x) = \int_\Lambda |x - a(\lambda)|^{-2} G^w \left(\lambda, \frac{x - a(\lambda)}{|x - a(\lambda)|}\right) d\lambda ,$$

$$G^w(\lambda, \omega) = -\frac{1}{8\pi^2} \int_{S^2} \delta'(\omega \cdot \theta) G(\lambda, \theta) |a'(\lambda) \cdot \theta| M(\theta, \lambda) d\theta, \ \omega \in S^2 .$$

This is the formula of Clack and Defrise (1994) and Kudo and Saito (1994). For more recent work see Katsevich (2002).

There exist various approximate solutions to the cone-beam reconstruction problem, most prominently the FDK formula for a circular source curve (which doesn't suffice Tuy's condition). It can be viewed as a clever adaption of the 2D filtered backprojection algorithm to 3D; see Feldkamp, Davis and Kress (1984). An approximate algorithm based on Orlov's condition of Theorem 2.7 is the π method of Danielsson et al. (1999).

References

1. M. Abramowitz and I.A. Stegun (1970): *Handbook of Mathematical Functions.* Dover.
2. G. Beylkin (1995): 'On the fast Fourier transform of functions with singularities', *Appl. Comp. Harm. Anal.* **2**, 363-381.
3. Y. Censor (1981): 'Row–action methods for huge and sparce systems and their applications', *SIAM Review* **23**, 444-466.
4. R. Clack and M. Defrise (1994): 'A Cone–Beam Reconstruction Algorithm Using Shift–Variant Filtering and Cone–Beam Backprojection', *IEEE Trans. Med. Imag.* **13**, 186-195.
5. P.E. Danielsson, P. Edholm, J. Eriksson, M. Magnusson Seger, and H. Turbell (1999): The original π-method for helical cone-beam CT, Proc. Int. Meeting on Fully 3D-reconstruction, Egmond aan Zee, June 3-6.
6. L.A. Feldkamp, L.C. Davis, and J.W. Kress (1984): 'Practical cone–beam algorithm', *J. Opt. Soc. Amer. A* **6**, 612-619.
7. K. Fourmont (1999): 'Schnelle Fourier–Transformation bei nicht–äquidistanten Gittern und tomographische Anwendungen', Dissertation *Fachbereich Mathematik und Informatik der Universität Münster, Münster, Germany.*
8. I.M. Gel'fand, M.I. Graev, N.Y. Vilenkin (1965): *Generalized Functions, Vol. 5: Integral Geometry and Representation Theory.* Academic Press.
9. P. Grangeat (1991): 'Mathematical framework of cone–beam reconstruction via the first derivative of the Radon transform' in: *G.T. Herman, A.K. Louis and F. Natterer (eds.): Lecture Notes in Mathematics* **1497**, 66-97.
10. C. Hamaker, K.T. Smith, D.C. Solmon and Wagner, S.L. (1980): 'The divergent beam X-ray transform', *Rocky Mountain J. Math.* **10**, 253-283.

11. C. Hamaker and D.C. Solmon (1978): 'The angles between the null spaces of X-rays', *J. Math. Anal. Appl.* **62**, 1-23.
12. S. Helgason (1999): *The Radon Transform.* Second Edition. Birkhäuser, Boston.
13. G.T. Herman (1980): *Image Reconstruction from Projection. The Fundamentals of Computerized Tomography.* Academic Press.
14. G.T. Herman and L. Meyer (1993): 'Algebraic reconstruction techniques can be made computationally efficient', *IEEE Trans. Med. Imag.* **12**, 600-609.
15. H.M. Hudson and R.S. Larkin (1994): 'Accelerated EM reconstruction using ordered subsets of projection data', *IEEE Trans. Med. Imag.* **13**, 601-609.
16. A.C. Kak and M. Slaney (1987): *Principles of Computerized Tomography Imaging.* IEEE Press, New York.
17. A. Katsevich (2002): 'Theoretically exact filtered backprojection-type inversion algorithm for spiral CT', *SIAM J. Appl. Math.* **62**, 2012-2026.
18. H. Kudo and T. Saito (1994): 'Derivation and implementation of a cone–beam reconstruction algorithm for nonplanar orbits', *IEEE Trans. Med. Imag.* **13**, 196-211.
19. F. Natterer (1986): *The Mathematics of Computerized Tomography.* John Wiley & Sons and B.G. Teubner. Reprint SIAM 2001.
20. F. Natterer and F. Wübbeling (2001): *Mathematical Methods in Image Reconstruction.* SIAM, Philadelphia.
21. F. Natterer and E.L. Ritman (2002): 'Past and Future Directions in X-Ray Computed Tomography (CT)', to appear in *Int. J. Imaging Systems & Technology.*
22. R.G. Novikov (2000): 'An inversion formula for the attenuated X-ray transform', *Preprint, Departement de Mathématique, Université de Nantes.*
23. V.A. Sharafutdinov (1994): *Integral geometry of Tensor Fields.* VSP, Utrecht.
24. G. Steidl (1998): 'A note on fast Fourier transforms for nonequispaced grids', *Advances in Computational Mathematics* **9**, 337-352.
25. S. Webb (1990): *From the Watching of Shadows.* Adam Hilger.

Adjoint Fields and Sensitivities for 3D Electromagnetic Imaging in Isotropic and Anisotropic Media

Oliver Dorn[1], Hugo Bertete-Aguirre[2], and George C. Papanicolaou[3]

[1] Gregorio Millán Institute of Fluid Dynamics, Nanoscience and Industrial
   Mathematics, Universidad Carlos III de Madrid, Avda. de la Universidad 30,
   28911 Leganés, Madrid, Spain
   odorn@math.uc3m.es
[2] Department of Mathematics, University of California Irvine, Irvine, CA, USA
   hbae@math.uci.edu
[3] Department of Mathematics, Building 380, 383V, Stanford University, Stanford,
   CA, USA
   papanico@math.stanford.edu

Summary. In this paper we give an overview of a recently developed method for solving an inverse Maxwell problem in environmental and geophysical imaging. Our main focus is on low-frequency cross-borehole electromagnetic induction tomography (EMIT), although related problems arise also in other applications in nondestructive testing and medical imaging. In typical applications (e.g. in environmental remediation), the isotropic or anisotropic conductivity distribution in the earth needs to be reconstructed from surface-to-borehole electromagnetic data. Our method uses a back-propagation strategy (based on adjoint fields) for solving this inverse problem. The method works iteratively, and can be considered as a nonlinear generalization of the Algebraic Reconstruction Technique (ART) in X-ray tomography, or as a nonlinear Kaczmarz-type approach. We will also propose a new regularization scheme for this method which is based on a proper choice of the function spaces for the inversion. A detailed sensitivity analysis for this problem is given, and a set of numerically calculated sensitivity functions for homogeneous isotropic media is presented.

1 Introduction: Electromagnetic Imaging of the Earth

The solid earth and certain pore fluids (such as hydrocarbons) are comparatively poor electrical conductors, while other pore fluids (such as brines) are good electrical conductors. If we wish to image the shape of conducting fluid plumes underground (as we might, e.g. for environmental cleanup purposes), one practical way of doing so is to make either electrical or electromagnetic measurements between boreholes (when two or more boreholes are available) or between borehole and surface arrays. One popular electromagnetic method of imaging without the need for boreholes is Ground Penetrating Radar (GPR), see for example Davis and Annan [DA89] or Fisher et al. [FMA92]. Here the probing waves are normally in the radio to microwave range of frequencies. GPR methods for imaging conductivity use comparatively high frequencies ($\geq 100\,\text{MHz}$) in order to have a signal that propagates through the Earth as a wave. The main practical difficulty with GPR imaging is that at these frequencies its depth of penetration is usually no deeper than about 5 m into the Earth in dry soil/sand, while maximum depths of less than 1–2 m are more typical in wet earth due to high local wave attenuation.

ElectroMagnetic Induction Tomography (EMIT) has been investigated recently as a promising new tool for imaging electrical conductivity variations in the earth (Wilt et al. [WAMBL, WMBTL] and Buettner et al. [BB99]) with a depth of investigation greater than Ground Penetrating Radar (GPR). The source is a magnetic field generated by currents in wire coils. This source field is normally produced in one borehole, while the received signals are the measured small changes in magnetic field in another, more distant borehole (Fig. 1). The method may also be used successfully in combination with surface sources and receivers. The goal of this procedure is to image electrical conductivity variations in the earth, much as X-ray tomography is used to image density variations through cross-sections of the body. However, the fields tend to diffuse rather than propagate as waves in this frequency band (from about 1 to 20 kHz) and the inversion techniques are therefore very different from those for GPR. Although field techniques have been developed and applied previously to collection of such EM data, the algorithms for inverting the magnetic field data to produce the desired images of electrical conductivity have not kept pace. The current state of the art in electromagnetic data inversion (Alumbaugh and Morrison [AM95a, AM95b]) is based on the Born/Rytov approximation (requiring

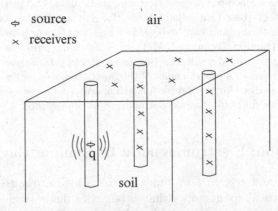

Fig. 1. The figure shows a possible experimental setup for 3D-EMIT. Three boreholes surround the area of interest. A z-directed dipole transmitter **q** is deployed in one of the boreholes, and data are gathered in the other boreholes and at the surface at positions $\mathbf{d}_n$, $n = 1, \ldots, N$. From these data, we want to recover the (isotropic or anisotropic) conductivity and permittivity distribution in the earth

a low contrast assumption), even though it is known that conductivity variations range over several orders of magnitude and therefore require nonlinear analysis. Other work on inversion for this problem has been presented for example by Habashy et al. [HGS93], Newman and Alumbaugh [NA97], and Haber et al. [HAO00]. An alternative fully nonlinear approach to solving the EMIT imaging problem was introduced recently by Dorn et al. [DBBP1, DBBP2]. The current paper is mainly concentrating on this inversion technique. The ultimate goal of our work is to produce better images of electrically conducting fluids underground by using robust inversion techniques over greater distances and higher contrasts than were previously possible.

Although other technologies, such as Electrical Resistance Tomography or ERT (Berryman and Kohn [BK90]; Daily et al. [DRLN]; Ramirez et al. [RD-LOC]; Borcea et al. [BBP96]; Borcea et al. [BBP99]) can produce electrical conductivity images at a useful spatial scale, the advantage of EMIT is that we can make use of existing monitoring wells and the surface to do our imaging. Since signals are transmitted and received inductively, we do not need to make ground contact (no ground penetrating electrodes); the technology is therefore relatively noninvasive. Furthermore, all conductors present can in principle be imaged using the EMIT technique, whereas ERT can only image those conductors maintaining a continuous electrical current path between the source and/or receiver electrodes. There is also the important potential advantage for both techniques that multiple frequencies can be employed to improve the imaging capability.

Our inversion method is based on the so-called 'adjoint technique', which (for the present application) has the very useful property that the inverse problem can be solved approximately by making two uses of the same forward modelling code. Using a somewhat oversimplified description of our technique, the updates to the electrical conductivity are obtained by first making one pass through the code using the latest best guess of the nature of the conducting medium, and then another pass with the adjoint operator applied to the differences in computed and measured data. Then the results of these two calculations are combined to determine updates to the original conductivity model. The resulting procedure is iterative and can be applied successively to parts of the data, e.g. data associated with one transmitter location can be used to update the model before other transmitter locations are considered. This procedure has several of the same advantages as wave equation migration in reflection seismology (Claerbout [Cla76]) and is also related to recent methods in electromagnetic migration introduced by Zhdanov et al. [ZTB96].

An important and interesting question that is commonly raised about all electrical imaging methods in the diffusive regime (frequencies $f < 1\,\mathrm{MHz}$) concerns the issue of *resolution*. Our physical ideas of resolution are normally based on concepts such as that of *Rayleigh criterion* for resolving power of an aperture in which two point sources of radiation are considered to be resolvable if the maximum of a diffraction pattern of one source coincides with the first minimum of the other (see Jenkins and White [JW57]). The spatial separation of two objects that can be resolved then clearly depends on the ratio of the wavelength of the radiating sources and the size of the aperture. When we are considering electromagnetic fields in the diffusive regime, it becomes problematic to define a useful resolution criterion. At very low frequencies such as those used in Electrical Resistance/Impedance Tomography (ERT/EIT) there is no wavelength associated with the signals and the resolution must instead be defined in terms of the numbers of source/receiver electrode pairs and the size of the region being imaged (see Borcea [Bor02] or Ramirez et al. [RDLOC]). The mathematically analogous problem in hydrology has also been treated (see Vasco et al. [VDL97]). For EMIT, the situation is closer to that of ERT/EIT than it is to that of wave propagation problem like seismic, ground penetrating radar, or optics. The wavelength does not provide a useful measure of the method's true resolution. A better approach in this situation is to define *sensitivity functions* (see for example Spies and Habashy [SH95]). A short introduction into sensitivity analysis for inverse problems is given in the following.

2 Sensitivities and the Adjoint Method

Sensitivity relations for inverse problems (or, more generally, parameter estimation problems) have been discussed many times in the literature. For example, a well-known reconstruction method in medical imaging is the filtered backprojection method in X-ray tomography [Nat86], which can be interpreted as a backprojection of filtered X-ray data of the human body along sensitivity functions. These sensitivity functions are concentrated on straight lines connecting the source and receiver positions, and the backprojection takes place uniformly over these lines. A generalization of this procedure can be found in Diffuse Optical Tomography (DOT) [Ar99], where the straight lines are replaced by 'banana-shaped' regions, which are in fact sensitivity functions distributed over the whole imaging domain with space-dependent weights. These weights are highest in some (banana-shaped) regions connecting the source and the receiver positions, but have nonvanishing values also in more remote areas of the imaging domain. For more details see for example Arridge et al. [Ar95a, Ar95b, Ar99] or Dorn [Dor00]. Both, the limited resolution usually encountered in the inverse problem of DOT and the severe ill-posedness are directly related to the specific shapes of these sensitivity functions.

The situation in low-frequency electromagnetic imaging is in many ways similar to the DOT problem. The physical propagation of signals is in both cases diffusive, which lets us expect that similar concepts of resolution should apply. The sensitivity functions in low-frequency EMIT are, as in DOT, distributed over the whole imaging domain. However, from energy arguments it follows that sensitivities far away from sources and receivers are small.

There are also some characteristic differences between DOT and EMIT. For example, in electromagnetic cross-borehole tomography, the sources and receivers are typically not located at the boundaries of the imaging domain, such that sidelobes of the sensitivity functions, which occur beyond the region of interest, have to be taken into account in the imaging process. Moreover, the propagating fields are now vector fields rather than scalar fields. This gives us a richer variety of sensitivity functions, depending on the component of the field which is used for collecting the data. We will see that the shapes of the sensitivity functions in EMIT show more variations than in DOT. For certain source-receiver geometries, the sensitivities in the region of interest are also here mainly concentrated on some almost 'banana-shaped' regions, whereas for other source-receiver constellations these regions look more like 'elongated doughnuts', with almost vanishing sensitivities close to the connecting line of source and receiver. We mention that another related geophysical application, which EMIT and DOT are interesting to compare with, is hydrology. That situation is treated in more details for example in Vasco et al. [VDL97].

The *nonlinear Kaczmarz-type approach* for solving the inversion problem (which can be characterized as a 'single-step adjoint-field inversion method') will be presented further below in this paper. It can be viewed as a nonlinear generalization of the algebraic reconstruction technique (ART) in X-ray tomography [Nat86,NW01]. In contrary to the filtered backprojection method mentioned earlier, ART uses only part of the data in each step of the reconstruction. It is an iterative technique and applies successive corrections to some initial guess for the parameter distribution, which in X-ray tomography is the *attenuation*. Each of these corrections is calculated by 'backprojecting' a (preprocessed or filtered) part of the residuals along those straight lines into the imaging domain which correspond to the incoming direction of the probing X-rays. We will show that, in an analogous way, each individual backprojection step of our generalized ART algorithm for EMIT takes place along sensitivity functions. An important difference of our generalized ART scheme to the classical ART algorithm is its nonlinearity. Whereas the sensitivities in the classical ART method are always concentrated on the same straight lines, the shapes of our sensitivity functions depend on the latest best guess for the parameters and therefore change with each update. The final reconstruction of our inversion method is given as a superposition of sensitivity functions with weights determined by the residual values which appear during the reconstruction process. The knowledge of the general structure of the sensitivity functions during the reconstruction process provides us with useful information about the resolution which we can expect from the final reconstruction.

In this paper we make use of the so-called *adjoint technique* for calculating sensitivity functions. In the more traditional 'direct method' for calculating sensitivities, typically as many forward problems need to be solved as there are parameters to be reconstructed. In large scale inverse problems these are typically several thousand parameters, one for each pixel or voxel used for the forward modelling. It has been pointed out in the literature that in such large scale parameter estimation problems the adjoint method can be much less computationally expensive than this 'direct method' for calculating sensitivities [Ar95a, Ar95b, AP98, Dor00, FO96, HAO00, MO90, MOEH, Nat96]. This is in particular true in cases where the Green's functions of the underlying forward problem (for example Maxwell's equations) are not known analytically,

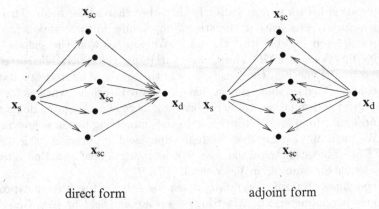

direct form   adjoint form

Fig. 2. Feynman diagrams visualizing direct and adjoint forms of sensitivity functions. $\mathbf{x}_s$ source position, $\mathbf{x}_d$ receiver position, $\mathbf{x}_{sc}$ scattering points

and have to be evaluated by employing some numerical forward modelling code. For example, given a source and a receiver position for which we want to calculate the space-dependent sensitivity functions in a 3D EMIT inversion problem, the 'direct method' would require us to solve 10^5–10^6 forward problems in order to find one sensitivity function. Using the 'adjoint method' we only need to solve one forward and one adjoint problem in order to calculate the same sensitivity function. The Feynman diagrams in Fig. 2 visualize this relationship. Each knot in the figure where one or more arrows *originate* indicates a separate run of the forward modelling code. Therefore, in the simple situation displayed in Fig. 2, we would have to solve six forward problems in order to calculate one sensitivity function in the direct form, whereas the adjoint scheme only requires two runs of our forward modelling code.

Sensitivity functions are closely related to Fréchet derivatives, which are evaluated at some reference parameter distribution. They describe the amount of which the receiver measurements will change if we slightly perturb the given reference parameter distribution at a certain position in the medium. Our linearized residual operator is just the Fréchet derivative of the nonlinear residual operator. We want to mention, however, that actually proving Fréchet differentiability of the nonlinear residual operator is not an easy task, and is beyond the scope of this overview. A possible method of proving Fréchet differentiability for general bilinear inverse problems can be found in [DDNPS]. We also mention that Fréchet differentiability for a related 3D-Maxwell problem has been discussed recently in [AB01].

Assuming that our residual operator is differentiable, it can be shown (and we will do that further below) that the sensitivity functions decompose this linearized residual operator as well as its adjoint operator. In many applications, in particular if the starting point is the discretized model of the forward problem, this Fréchet derivative is called the 'Jacobian' or 'sensitivity matrix' of the underlying forward operator. The sensitivity functions can then be interpreted as the rows (or columns) of the Jacobian or sensitivity matrix. They play a major role in basically all gradient-based inversion methods which try to minimize the least squares measure of the misfit between given data and predicted data corresponding to a reference medium. Therefore, the ideas presented here are certainly applicable also beyond the single step adjoint field inversion method which we are concentrating on in this paper.

3 Maxwell's Equations for Anisotropic Media

The system of time-harmonic Maxwell's Equations in three spatial dimensions can be formulated as follows

$$\nabla \times \mathbf{H}(\mathbf{x}) - (i\omega\epsilon(\mathbf{x}) + \sigma(\mathbf{x}))\,\mathbf{E}(\mathbf{x}) = \mathbf{J}(\mathbf{x}), \tag{1}$$

$$\nabla \times \mathbf{E}(\mathbf{x}) + i\omega\mu(\mathbf{x})\,\mathbf{H}(\mathbf{x}) = \mathbf{M}(\mathbf{x}), \tag{2}$$

$$\nabla \cdot \epsilon\mathbf{E}(\mathbf{x}) = 0, \qquad \nabla \cdot \mu\mathbf{H}(\mathbf{x}) = 0, \tag{3}$$

where $\mathbf{E}(\mathbf{x})$ and $\mathbf{H}(\mathbf{x})$ denote the electric and magnetic field, $\mathbf{J}(\mathbf{x})$ and $\mathbf{M}(\mathbf{x})$ are the electric and magnetic source currents, $\omega > 0$ is the angular frequency, $\omega = 2\pi f$, and $i = \sqrt{-1}$. The parameters describing the underlying medium are the electric conductivity $\sigma(\mathbf{x})$ with dimension Siemens per metre ($\mathrm{S\,m^{-1}}$), the dielectric permittivity $\epsilon(\mathbf{x})$ with dimension Farad per metre ($\mathrm{F\,m^{-1}}$), and the magnetic permeability $\mu(\mathbf{x})$ with dimension Henry per metre ($\mathrm{H\,m^{-1}}$). We assume that all these parameters are strictly positive in the physical domain $\sigma(\mathbf{x}) > 0$, $\epsilon(\mathbf{x}) > 0$, and $\mu(\mathbf{x}) > 0$ in $\mathbb{R}^3$. Equations (3) indicate that there are no electric or magnetic space charges in the medium.

In the case of isotropic media the physical parameters are described by (complex valued) scalar functions. For anisotropic media these have to be replaced by general 3×3 tensor functions. We will assume here without loss of generality that $\sigma(\mathbf{x})$, $\epsilon(\mathbf{x})$, and $\mu(\mathbf{x})$ are described by diagonal tensors with real positive numbers on the diagonal and zero off-diagonal elements [DBBP2]. We get the representations

$$\begin{aligned}
\sigma(\mathbf{x}) &= \mathbf{diag}(\sigma^{(x)}(\mathbf{x}), \sigma^{(y)}(\mathbf{x}), \sigma^{(z)}(\mathbf{x})), \\
\epsilon(\mathbf{x}) &= \mathbf{diag}(\epsilon^{(x)}(\mathbf{x}), \epsilon^{(y)}(\mathbf{x}), \epsilon^{(z)}(\mathbf{x})), \\
\mu(\mathbf{x}) &= \mathbf{diag}(\mu^{(x)}(\mathbf{x}), \mu^{(y)}(\mathbf{x}), \mu^{(z)}(\mathbf{x})).
\end{aligned} \tag{4}$$

For the following, we introduce the notation

$$a = i\omega\mu, \qquad b = i\omega\epsilon + \sigma, \tag{5}$$

with

$$\begin{aligned}
a(\mathbf{x}) &= \mathbf{diag}(a^{(x)}(\mathbf{x}), a^{(y)}(\mathbf{x}), a^{(z)}(\mathbf{x})), \\
b(\mathbf{x}) &= \mathbf{diag}(b^{(x)}(\mathbf{x}), b^{(y)}(\mathbf{x}), b^{(z)}(\mathbf{x})).
\end{aligned}$$

With the notation of (5) we write (1), (2) as

$$\nabla \times \mathbf{H}(\mathbf{x}) - b(\mathbf{x})\,\mathbf{E}(\mathbf{x}) = \mathbf{J}(\mathbf{x}), \tag{6}$$
$$\nabla \times \mathbf{E}(\mathbf{x}) + a(\mathbf{x})\,\mathbf{H}(\mathbf{x}) = \mathbf{M}(\mathbf{x}). \tag{7}$$

This is the form of Maxwell's equations that we will use.

Equations (6), (7) describe the propagation of an electromagnetic field which is generated by a (time-harmonic) source distribution

$$\tilde{\mathbf{q}}(\mathbf{x}, t) = \mathbf{q}(\mathbf{x})e^{i\omega t} = \begin{pmatrix} \mathbf{J}(x) \\ \mathbf{M}(x) \end{pmatrix} e^{i\omega t},$$

where we typically will apply a magnetic dipole source of the form

$$\mathbf{q}(\mathbf{x}) = \begin{pmatrix} \mathbf{0} \\ \mathbf{m_s}\delta(\mathbf{x} - \mathbf{s}) \end{pmatrix}. \tag{8}$$

Here, $\mathbf{m}_s$ is the magnetic dipole moment of the source and $\mathbf{s}$ the source position. In the following we consider (6), (7) in the unbounded domain $\Omega = \mathbb{R}^3$ for a given source (8). In Sect. 6, we will extend the analysis to the case when data corresponding to many source positions are used.

We assume that (6), (7) provide a model for the propagation of the electric and magnetic field emitted by the source $\mathbf{q}$, and that we measure the magnetic field $\mathbf{H}(\mathbf{d}_n)$ for N discrete detector positions $\mathbf{d}_n$, $n = 1, \ldots, N$.

In geophysical applications, it is often the case that the coefficient a does not vary significantly in the region of interest. In this paper, we will therefore assume that a is known in the whole domain Ω, e.g. from interpolation of well-log data. In our numerical experiments, we will furthermore only consider the case that a is constant everywhere, although the discussion of the paper applies to the more general situation of space dependent parameters $a(x)$. The coefficient b is assumed to be unknown in the medium and has to be recovered from the data. The inverse problem we consider here is therefore as follows.

Inverse Problem: Assume that for many source distributions $\mathbf{q}_j = (\mathbf{J}_j, \mathbf{M}_j)^T$, $j = 1, \ldots, p$, for example of the form (8), the corresponding magnetic fields $\mathbf{H}_j(\mathbf{d}_n)$ are measured at the receiver locations $\mathbf{d}_n$, $n = 1 \ldots, N$. Find a parameter distribution $\tilde{b}$ such that for the solutions $(\tilde{\mathbf{E}}_j, \tilde{\mathbf{H}}_j)^T$ of

$$\nabla \times \tilde{\mathbf{H}}_j - \tilde{b}\,\tilde{\mathbf{E}}_j = \mathbf{J}_j, \tag{9}$$
$$\nabla \times \tilde{\mathbf{E}}_j + a\,\tilde{\mathbf{H}}_j = \mathbf{M}_j, \tag{10}$$

the magnetic field values at the receiver positions coincide with the data.

If there is no noise in the data, then there may be many parameter distributions $\tilde{b}$ for (9), (10) that satisfy the data, so we have to select one using a regularization criterion. If there is noise, the usual case, then we need to minimize the sum of the squares of the differences between the data calculated by (9), (10) and those which are physically measured.

4 The Linearized Residual Operator

We want to define now the basic function spaces which are used throughout this paper. The main objective of this section is to introduce the inner products which will be used in the derivation of the adjoint of the linearized residual operator which is discussed in Sect. 5.1. For simplicity in the notation, we will assume in the present section that we have only one source function $\mathbf{q}(\mathbf{x})$ given which is fixed. In Sect. 6 we will come back to the situation where more than one physical experiment are performed for gathering the data.

The basic spaces which we will need are as follows.

$$F = \left\{ b = \mathbf{diag}\left(b^{(x)}, b^{(y)}, b^{(z)} \right) \; ; \; b^{(x)}, b^{(y)}, b^{(z)} \in L^2(\Omega) \right\},$$

$$Z = \left\{ \zeta = (\mathbf{h}_1, \ldots, \mathbf{h}_N), \quad \mathbf{h}_n \in \mathbb{C}^3 \right\},$$

$$U = \left\{ \begin{pmatrix} \mathbf{E}(\mathbf{x}) \\ \mathbf{H}(\mathbf{x}) \end{pmatrix} \; ; \; \mathbf{E}, \mathbf{H} \in [L^2(\Omega)]^3 \right\},$$

$$Y = \left\{ \begin{pmatrix} \mathbf{J}(\mathbf{x}) \\ \mathbf{M}(\mathbf{x}) \end{pmatrix} \; ; \; \mathbf{J}, \mathbf{M} \in [L^2(\Omega)]^3 \right\}.$$

Notice that in the case of isotropic media we have $b^{(x)} = b^{(y)} = b^{(z)} = b$ such that we typically will identify in that situation $F = L^2(\Omega)$ with $b \in F$. It will always be clear from the context which space we refer to, such that we omit using different symbols for this space in the cases of isotropic and anisotropic media. All these quantities carry a physical dimension, which has to be taken into account when introducing the following inner products for these spaces.

$$\left\langle \begin{pmatrix} \mathbf{E}_1 \\ \mathbf{H}_1 \end{pmatrix}, \begin{pmatrix} \mathbf{E}_2 \\ \mathbf{H}_2 \end{pmatrix} \right\rangle_U = \int_\Omega \gamma_E^2 \mathbf{E}_1 \overline{\mathbf{E}_2}\, d\mathbf{x} + \int_\Omega \gamma_H^2 \mathbf{H}_1 \overline{\mathbf{H}_2}\, d\mathbf{x},$$

$$\left\langle \begin{pmatrix} \mathbf{J}_1 \\ \mathbf{M}_1 \end{pmatrix}, \begin{pmatrix} \mathbf{J}_2 \\ \mathbf{M}_2 \end{pmatrix} \right\rangle_Y = \int_\Omega \gamma_J^2 \mathbf{J}_1 \overline{\mathbf{J}_2}\, d\mathbf{x} + \int_\Omega \gamma_M^2 \mathbf{M}_1 \overline{\mathbf{M}_2}\, d\mathbf{x},$$

$$\langle \zeta_1, \zeta_2 \rangle_Z = \sum_{n=1}^N \gamma_H^2 \mathbf{h}_{1,n} \overline{\mathbf{h}_{2,n}} = \sum_{n=1}^N \sum_{i=x,y,z} \gamma_H^2 h_{1,n}^{(i)} \overline{h_{2,n}^{(i)}}$$

with $\zeta_l = (\mathbf{h}_{l,1}, \ldots, \mathbf{h}_{l,N})$, $\hat{\mathbf{h}}_{l,n} := \gamma_H \mathbf{h}_{l,n} \in \mathbb{C}^3$ for $l = 1, 2$, $n = 1, \ldots, N$.

$$\left\langle \mathbf{diag}(b_1^{(x)}, b_1^{(y)}, b_1^{(z)}), \mathbf{diag}(b_2^{(x)}, b_2^{(y)}, b_2^{(z)}) \right\rangle_F$$

$$= \int_\Omega \gamma_b^2 \left(b_1^{(x)}(\mathbf{x}) \overline{b_2^{(x)}(\mathbf{x})} + b_1^{(y)}(\mathbf{x}) \overline{b_2^{(y)}(\mathbf{x})} + b_1^{(z)}(\mathbf{x}) \overline{b_2^{(z)}(\mathbf{x})} \right) d\mathbf{x}.$$

An analogous expression holds for the inner product of $F = L^2(\Omega)$ in the case of isotropic media. The constants γ_E, γ_H, γ_J, γ_M and γ_b are introduced in order to make the given scalar products dimensionless. More precisely, we have $\gamma_E = [\mathbf{E}]^{-1}$, $\gamma_H = [\mathbf{H}]^{-1}$, $\gamma_J = [\mathbf{J}]^{-1}$, $\gamma_M = [\mathbf{M}]^{-1}$, and $\gamma_b = [b]^{-1}$, where the symbol [.] denotes the dimension of the corresponding physical quantity. Notice that $\gamma_b \gamma_E = \gamma_J$, which follows immediately from (6). We will also consider in this paper the corresponding dimensionless quantities which we always indicate by a roof on top of the symbol, for example $\hat{b} = \gamma_b b$ for $b \in F$ or $\hat{h} = \gamma_H h$ for $h \in Z$. For these quantities, we will accordingly use the inner products $\langle , \rangle_{\hat{F}}$, $\langle , \rangle_{\hat{Z}}, \ldots$, which are defined as above but without the additional factors in the integrals. The spaces F, U, Y and Z (as well as $\hat{F}$, $\hat{U}$, $\hat{Y}$ and $\hat{Z}$) are Hilbert spaces with these inner products.

We write (6), (7) in system form

$$\Lambda_M(b)\, \mathbf{u} = \mathbf{q}, \tag{11}$$

with

$$\Lambda_M(b) = \begin{pmatrix} -b \ \nabla\times \\ \\ \nabla\times \ a \end{pmatrix}, \tag{12}$$

and

$$\mathbf{u} = \begin{pmatrix} \mathbf{E} \\ \mathbf{H} \end{pmatrix}, \qquad \mathbf{q} = \begin{pmatrix} \mathbf{J} \\ \mathbf{M} \end{pmatrix}.$$

The Maxwell Operator $\Lambda_M(b)$ is defined on a dense subset of the Hilbert space U, given the parameter distribution b. When the conductivity is everywhere positive, $\sigma > 0$, (11) has a unique solution $\mathbf{u} \in U$ for each source $\mathbf{q} \in Y$. For a function $\mathbf{u} \in U$ of (11) we define the *measurement operator* $M : U \to Z$ by

$$(M\mathbf{u})_n(\mathbf{x}) = \int_\Omega \delta(\mathbf{x} - \mathbf{d}_n)\, \mathbf{H}(\mathbf{x})\, d\mathbf{x} \qquad (n = 1, \ldots, N) \tag{13}$$

with $\mathbf{u} = \begin{pmatrix} \mathbf{E} \\ \mathbf{H} \end{pmatrix}$, and with $\mathbf{d}_n$, $n = 1, \ldots, N$, being the detector positions. The measurement operator M is well defined by (13) and maps the fields $\mathbf{u} \in U$ to a vector of complex numbers $M\mathbf{u} \in Z$, which consists of point measurements of the magnetic field components at the detector positions $\mathbf{d}_n$, $n = 1, \ldots, N$. More precisely, we can write

$$M\begin{pmatrix} \mathbf{E} \\ \mathbf{H} \end{pmatrix} = \left(\mathbf{H}(\mathbf{d}_1),\, \mathbf{H}(\mathbf{d}_2),\, \ldots,\, \mathbf{H}(\mathbf{d}_N) \right)^T,$$

where the symbol ^{T} stands for transpose. For the given source $\mathbf{q}$ and given data $g \in Z$ we define the *residual operator* R by

$$R : F \to Z, \qquad R(b) = M\mathbf{u}(b) - g,$$

where $\mathbf{u}(b)$ is the solution of (11) with parameter b. The operator R is a *nonlinear* operator from the space of parameters F to the space of measurements Z.

We will assume that the source $\mathbf{q}$ is given by (8) with

$$\mathbf{m}_s = \alpha_1 \mathbf{e}_1 + \alpha_2 \mathbf{e}_2 + \alpha_3 \mathbf{e}_3, \tag{14}$$

with $\hat{\alpha}_i := \gamma_M \alpha_i \in \mathbb{C}$ for $i = 1, 2, 3$. The symbols $\mathbf{e}_i$, $i = 1, 2, 3$, will always denote the cartesian unit vectors, for example $\mathbf{e}_1 = (1, 0, 0)^T$. We will also often replace $i = 1, 2, 3$ by $i = x, y, z$ in the notation in order to emphasize the spatial dimension which we refer to. We denote by $\mathbf{E}(\mathbf{x}) = (E^{(x)}(\mathbf{x}), E^{(y)}(\mathbf{x}), E^{(z)}(\mathbf{x}))^T$ (and similarly $\mathbf{H}(\mathbf{x})$) the solution of the *forward problem*

$$\Lambda_M(b) \begin{pmatrix} \mathbf{E} \\ \mathbf{H} \end{pmatrix} = \begin{pmatrix} 0 \\ \mathbf{m}_s \delta(\mathbf{x} - \mathbf{s}) \end{pmatrix} \tag{15}$$

with $\Lambda_M(b)$ defined as in (12), and a, b defined as in (5). In order to find a useful representation of the *linearized residual operator* R' of R, we perturb b, $\mathbf{E}$ and $\mathbf{H}$ by

$$b \rightarrow b + \delta b \qquad \mathbf{E} \rightarrow \mathbf{E} + \mathbf{v} \qquad \mathbf{H} \rightarrow \mathbf{H} + \mathbf{w},$$

and plug this into (6), (7). When neglecting terms of higher than linear order in δb, v and w, and applying the measurement operator M, we arrive at the following result.

Theorem 4.1. *The linearized residual operator $R'[b]$ is given by*

$$R'[b] \, \delta b = M \begin{pmatrix} \mathbf{v} \\ \mathbf{w} \end{pmatrix}, \tag{16}$$

where $\begin{pmatrix} \mathbf{v} \\ \mathbf{w} \end{pmatrix}$ solves

$$\Lambda_M(b) \begin{pmatrix} \mathbf{v} \\ \mathbf{w} \end{pmatrix} = \begin{pmatrix} \delta b \, \mathbf{E} \\ 0 \end{pmatrix}. \tag{17}$$

5 Adjoint Sensitivity Analysis for Anisotropic Media

5.1 The Adjoint Linearized Residual Operator

The *adjoint linearized residual operator* $R'[b]^*$ is formally defined by the identity

$$\left\langle R'[b] \, \delta b, \, \zeta \right\rangle_Z = \left\langle \delta b, \, R'[b]^* \zeta \right\rangle_F.$$

It is a mapping from the data space Z into the parameter space F. In the next theorem, we will present a useful expression for the action of $R'[b]^*$ on a given vector $\zeta \in Z$. The proof of this theorem can be found in [DBBP2].

Theorem 5.1. *Let us assume that we are given $\zeta = (\mathbf{h}_1, \ldots, \mathbf{h}_N) \in Z$ with $\hat{\mathbf{h}}_n = (\hat{h}_n^{(x)}, \hat{h}_n^{(y)}, \hat{h}_n^{(z)})^T \in \mathbb{C}^3$ for $n = 1, \ldots, N$. Furthermore, we denote by $\mathcal{E}(\mathbf{x}) = (\mathcal{E}^{(x)}(\mathbf{x}), \mathcal{E}^{(y)}(\mathbf{x}), \mathcal{E}^{(z)}(\mathbf{x}))^T$ the solution of the adjoint problem*

$$\Lambda_M^*(b)\begin{pmatrix}\mathcal{E}\\\mathcal{H}\end{pmatrix} = \begin{pmatrix}0\\\gamma_M^{-1}\sum_{n=1}^N\sum_{i=x,y,z}\hat{h}_n^{(i)}\mathbf{e}_i\delta(\mathbf{x}-\mathbf{d}_n)\end{pmatrix}, \tag{18}$$

with

$$\Lambda_M^*(b) = \begin{pmatrix}-\overline{b} & \nabla\times\\ \nabla\times & \overline{a}\end{pmatrix}, \tag{19}$$

$$\overline{b} = -i\omega\epsilon + \sigma, \quad \overline{a} = -i\omega\mu.$$

Then $R'[b]^*\zeta \in F$ is given by

$$\left(R'[b]^*\zeta\right)(\mathbf{x}) = \gamma_b^{-1}\mathbf{diag}\left(\overline{\hat{E}^{(x)}(\mathbf{x})}\hat{\mathcal{E}}^{(x)}(\mathbf{x}),\ \overline{\hat{E}^{(y)}(\mathbf{x})}\hat{\mathcal{E}}^{(y)}(\mathbf{x}),\ \overline{\hat{E}^{(z)}(\mathbf{x})}\hat{\mathcal{E}}^{(z)}(\mathbf{x})\right), \tag{20}$$

where $\mathbf{E}$ is the solution of (15).

5.2 Decomposition of the Residual Operator and Its Adjoint

In this section, we will present the basic decompositions of the linearized residual operator and its adjoint into sensitivity functions. The proofs of the theorems in this section and in the next section can be found in [DBBP2].

Theorem 5.2. *Given a source (14). We can find functions* $S_k^{(i)}(n,l;\mathbf{y})$, $i = 1,2,3$, *such that the linearized residual operator* $R'[b]$ *can be decomposed as*

$$\left(R'[b]\delta b\right)_n = \gamma_H^{-1}\sum_{l=1}^3\left(\sum_{i=1}^3\int_\Omega\widehat{\delta b}^{(i)}(\mathbf{y})\left(\sum_{k=1}^3\hat{\alpha}_k\ S_k^{(i)}(n,l;\mathbf{y})\right)d\mathbf{y}\right)\mathbf{e}_l. \tag{21}$$

Moreover, the same functions $S_k^{(i)}(n,l;\mathbf{y})$ *also decompose the adjoint linearized residual operator* $R'[b]^*$ *in the following way*

$$\left(R'[b]^*\zeta\right)^{(i)}(\mathbf{y}) = \gamma_b^{-1}\sum_{n=1}^N\sum_{l=1}^3\hat{h}_n^{(l)}\left(\sum_{k=1}^3\overline{\hat{\alpha}_k S_k^{(i)}(n,l;\mathbf{y})}\right). \tag{22}$$

The functions $S_k^{(i)}(n,l;\mathbf{y})$ *are called* anisotropic sensitivity functions.

In the following, we will first define these sensitivity functions $S_k^{(i)}(n,l;\mathbf{y})$, and then reformulate the above theorem in a slightly more formal way. First, we decompose the given vector $\zeta = (\mathbf{h}_1,\ldots,\mathbf{h}_N) \in Z$ as

$$\mathbf{h}_n = h_n^{(1)}\mathbf{e}_1 + h_n^{(2)}\mathbf{e}_2 + h_n^{(3)}\mathbf{e}_3 \tag{23}$$

with $\hat{h}_n^{(i)} := \gamma_H h_n^{(i)} \in \mathbb{C}$ for $i = 1,2,3$ and $n = 1,\ldots,N$. Let $(\mathbf{E}_k,\mathbf{H}_k)^T$, $k = 1,2,3$, be the solutions of the direct problems

$$\Lambda_M(b)\begin{pmatrix}\mathbf{E}_k\\\mathbf{H}_k\end{pmatrix} = \begin{pmatrix}0\\\gamma_M^{-1}\mathbf{e}_k\delta(\mathbf{x}-\mathbf{s})\end{pmatrix}. \tag{24}$$

Moreover, let $(\mathcal{E}_{n,l}, \mathcal{H}_{n,l})^T$, $l = 1, 2, 3$, $n = 1 \ldots, N$, be solutions of the adjoint problems

$$\Lambda_M^*(b) \begin{pmatrix} \mathcal{E}_{n,l} \\ \mathcal{H}_{n,l} \end{pmatrix} = \begin{pmatrix} 0 \\ \gamma_M^{-1} \mathbf{e}_l \delta(\mathbf{x} - \mathbf{d}_n) \end{pmatrix}, \tag{25}$$

where $\Lambda_M^*(b)$ is defined as in (19). With this notation, the *anisotropic sensitivity functions* $S_k^{(i)}(n, l; \mathbf{y})$ are defined by

$$S_k^{(i)}(n, l; \mathbf{y}) = \hat{E}_k^{(i)}(\mathbf{y}) \overline{\hat{\mathcal{E}}_{n,l}^{(i)}(\mathbf{y})}, \tag{26}$$

where the fields $(\mathbf{E}, \mathbf{H})^T$ are solutions of (15). Notice that these sensitivity functions are defined to be dimensionless.

Now, making use of the inner products which have been introduced in Sect. 4, we get the following equivalent formulation of Theorem 5.2.

Theorem 5.3. *Given a source (14). Define the functions* $S_k^{(i)}(n, l; \mathbf{y})$ *as in (26). Then we have the following decompositions of the linearized residual operator* $R'[b]$ *and its adjoint* $R'[b]^*$.
Projection Step:

$$\left(R'[b] \delta b \right)_n = \gamma_H^{-1} \sum_{l=1}^{3} \left(\sum_{i=1}^{3} \sum_{k=1}^{3} \hat{\alpha}_k \left\langle \widehat{\delta b}^{(i)} , \overline{S_k^{(i)}(n, l; \, . \,)} \right\rangle_{\hat{F}} \right) \mathbf{e}_l. \tag{27}$$

Backprojection Step:

$$\left(R'[b]^* \zeta \right)^{(i)} (\mathbf{y}) = \gamma_b^{-1} \sum_{k=1}^{3} \hat{\alpha}_k \left\langle \hat{\zeta}, S_k^{(i)}(\, . \, , \, . \, ; \mathbf{y}) \right\rangle_{\hat{Z}}. \tag{28}$$

Remark 5.1. We note that (27) gives rise to the interpretation of $R'[b]$ as a 'projection operator'. It 'projects' each component $\delta b^{(i)}$ of the parameter perturbation δb along the sensitivity function $S_k^{(i)}(n, l; \, . \,)$ into the data space Z. Summing up all these contributions yields the (linearized) residuals $R'[b] \delta b$. Analogously, (28) gives rise to the interpretation of $R'[b]^*$ as a 'backprojection operator'. It 'backprojects' the residual vector ζ of the data space Z along the sensitivity functions $S_k^{(i)}(\, . \, , \, . \, ; \mathbf{y})$ into the parameter space F.

5.3 A Special Case: Isotropic Media

Let us consider an isotropic medium, i.e. let $b_{\text{iso}}(\mathbf{y})$, $\delta b_{\text{iso}}(\mathbf{y})$ be given by

$$b_{\text{iso}}(\mathbf{y}) = \mathbf{diag} \left(b^{(0)}(\mathbf{y}), b^{(0)}(\mathbf{y}), b^{(0)}(\mathbf{y}) \right),$$

$$\delta b_{\text{iso}}(\mathbf{y}) = \mathbf{diag} \left(\delta b^{(0)}(\mathbf{y}), \delta b^{(0)}(\mathbf{y}), \delta b^{(0)}(\mathbf{y}) \right).$$

We can replace the diagonal tensors for describing the parameters δb_{iso} and b_{iso} by the scalar functions $\delta b^{(0)}$, $b^{(0)}$, which live in the reduced function space $L^2(\mathbb{R}^3)$. The *isotropic sensitivity functions* $S_k(n, l; \mathbf{y})$ are defined as

$$S_k(n, l; \mathbf{y}) = \hat{\mathbf{E}}_k(\mathbf{y}) \cdot \overline{\hat{\mathcal{E}}_{n,l}(\mathbf{y})}, \tag{29}$$

where $\mathbf{E}_k(\mathbf{y})$ solves (24), $\mathbf{E}(\mathbf{y})$ solves (15), and $\mathcal{E}_{n,l}(\mathbf{y})$ solves (25). Obviously, we have

$$S_k(n, l; \mathbf{y}) = \sum_{i=1}^{3} S_k^{(i)}(n, l; y). \tag{30}$$

With these functions, we can formulate the analogous form of Theorem 5.2 for isotropic media.

Theorem 5.4. *Given a source (14). Let the functions $S_k(n, l; \mathbf{y})$ be defined as in (29). Then we have the following decompositions of $R'[b]$*

$$\left(R'[b]\delta b\right)_n = \gamma_H^{-1} \sum_{l=1}^{3} \left(\sum_{k=1}^{3} \hat{\alpha}_k \int_{\Omega} \widehat{\delta b}(\mathbf{y}) S_k(n, l; \mathbf{y})\, d\mathbf{y}\right) \mathbf{e}_l. \tag{31}$$

In addition, $R'[b]^*$ can be decomposed as

$$\left(R'[b]^*\zeta\right)(\mathbf{y}) = \gamma_b^{-1} \sum_{n=1}^{N} \sum_{l=1}^{3} \hat{h}_n^{(l)} \sum_{k=1}^{3} \overline{\hat{\alpha}_k S_k(n, l; \mathbf{y})}. \tag{32}$$

Using a more formal notation, this theorem can be reformulated as follows.

Theorem 5.5. *Given a source (14). Define the functions $S_k(n, l; \mathbf{y})$ as in (29). Then, the linearized residual operator $R'[b]$ and its adjoint $R'[b]^*$ have the following decompositions.*
Projection step:

$$\left(R'[b]\delta b\right)_n = \gamma_H^{-1} \sum_{l=1}^{3} \sum_{k=1}^{3} \hat{\alpha}_k \left\langle \widehat{\delta b}, \overline{S_k(n, l; \cdot)} \right\rangle_{\hat{F}} \mathbf{e}_l. \tag{33}$$

Backprojection step:

$$\left(R'[b]^*\zeta\right)(\mathbf{y}) = \gamma_b^{-1} \sum_{k=1}^{3} \hat{\alpha}_k \left\langle \hat{\zeta}, S_k(\,.\,,\,.\,;\mathbf{y}) \right\rangle_{\hat{Z}}. \tag{34}$$

6 A Single Step Adjoint Field Inversion Scheme

6.1 Outline of the Inversion Scheme

In this section we will give a brief outline of the adjoint field inversion scheme for solving the 3D EMIT problem. More details can be found in [DBBP1, DBBP2]. It was already mentioned earlier that the method can be considered as a non-linear generalization of the algebraic reconstruction technique (ART) in X-ray

tomography [Nat86], and it is also similar to the simultaneous iterative reconstruction technique (SIRT) as presented in [DL79] or to the Kaczmarz method
for solving linear systems of equations (see also [Nat86, NW01, NW95, Dor98]).

We will assume for simplicity that the medium is isotropic. Extensions to the
case of anisotropic media are straightforward. When solving the inverse problem
which was formulated in Sect. 3, we ideally want to find a parameter $\tilde{b}$ such that
for all given source positions $\mathbf{q}_j$, $j = 1, \ldots, p$, the residuals vanish

$$R_j(\tilde{b}) = 0, \qquad j = 1, \ldots, p. \tag{35}$$

The index j refers here to the actual source position $\mathbf{q}_j$ which is considered by
the operator $R_j(\tilde{b})$. Using real data, we cannot be sure whether these equations
can be satisfied exactly. Therefore, we generalize our criterion for a solution.
Defining the least squares cost functionals

$$\mathcal{J}_j(b) = \frac{1}{2}\|R_j(b)\|_{L_2}^2 \tag{36}$$

for $j = 1, \ldots, p$, we are searching for a joint minimizer of these cost functionals.
A standard method for finding a minimizer of (36) is to start a search with some
initial guess $b^{(0)}$, and to find descent directions for the sum

$$\mathcal{J}(b) = \sum_{j=1}^{p} \mathcal{J}_j(b)$$

in each step of an iterative scheme. Popular choices for descent directions are
for example the gradient direction, conjugate gradient directions, Newton- or
Quasi-Newton directions (see for example [DS96, NW99, Vog02] for details). In
our Kaczmarz-type approach, however, we consider only individual terms of (36)
in a given step. We search for descent directions only with respect to the chosen
term $\mathcal{J}_j(b)$, and apply it to the latest best guess. In the next step, we will then
choose a different term $\mathcal{J}_j(b)$ of (36) and proceed in the same way. After having
used each of the terms in $\mathcal{J}_j(b)$ exactly once, we have completed one *sweep* of
our algorithm.

This strategy of switching back and forth between different data subsets
does not only have the advantage that a correction of our latest best guess
for the parameters can be achieved more rapidly, but it also helps to avoid
getting trapped in so-called 'local minima' of more classical gradient-based search
algorithms. This is because the variety of search criteria is increased in our
algorithm. Instead of having just one search direction available per data set
(considering only descent directions of $\mathcal{J}(b)$), we have now as many useful search
directions as we have source positions. Even if one of these search directions does
not advance our reconstruction, typically at least one of the remaining search
directions gives us an improvement of it. Notice also that the combined cost
$\mathcal{J}(b)$ is certainly allowed to increase for a while during the reconstruction using
the Kaczmarz-type scheme. This will for example happen when a gradient-based
inversion algorithm with respect to $\mathcal{J}(b)$ would get close to a local minimum,

but at least one of the individual Kaczmarz-type search directions is not affected by this local minimum of the combined least-squares cost functional.

We will in the following derive our Kaczmarz-type update directions for reducing the individual terms (36) using the formulation (35). These update directions will have the useful property that they can easily be generalized in order to incorporate efficient regularization schemes in our algorithm.

In order to find an 'update' (or 'correction') δb for our parameter b we linearize the nonlinear operator R_j (assuming that this linearized operator $R_j'[b]$ exists and is well-defined) and write

$$R_j(b + \delta b) = R_j(b) + R_j'[b]\delta b + O(||\delta b||^2). \tag{37}$$

The linearized operator $R_j'[b]$ is often called the Fréchet derivative of R_j at b. (See for example [DDNPS] and references therein for some formal derivations of Fréchet derivatives in different applications.) It is also closely related to the 'sensitivity functions' of the parameter profile with respect to the data. Using (37) we want to look for a correction δb such that $R_j(b + \delta b) = 0$. Neglecting terms of order $O(||\delta b||^2)$ in (37), this amounts to solving

$$R_j'[b]\delta b = -R_j(b) . \tag{38}$$

Certainly, due to the ill-posedness of our problem, this equation needs to be handled with care. Treated as an ill-posed linear inverse problem, a classical solution of (38) will be the minimum-norm solution

$$\delta b_{MN} = -R_j'[b]^* \left(R_j'[b]R_j'[b]^*\right)^{-1} R_j(b), \tag{39}$$

where $R_j'[b]^*$ is the adjoint operator of $R_j'[b]$ with respect to our chosen spaces F and Z. In applications with very few data, this form has the useful property that it avoids contributions in the solution which are in the (often non-empty) null-space of the (linearized) forward operator $R_j'[b]$. Using (37) it can be verified by direct calculation that

$$\mathcal{J}_j(b + \kappa\delta b_{MN}) = \mathcal{J}_j(b) - \kappa||R_j(b)||_Z^2 + O(||\kappa\delta b_{MN}||_F^2) \tag{40}$$

for a step-size $\kappa > 0$, such that (39) also is a descent direction of the least squares cost functional (36).

In our application the operator $C = (R_j'[b]R_j'[b]^*)^{-1}$ is very ill-conditioned, such that a regularized version needs to be used. This can be, for example, $\hat{C} = (R_j'[b]R_j'[b]^* + \lambda I)^{-1}$ where λ is some regularization parameter and I is the identity operator. Unfortunately, in practice both, C and $\hat{C}$, are very expensive to calculate and to apply to the residuals R_j. Typically, a direct calculation of the operator $\hat{C}$ would require us to solve as many forward and adjoint problems as we have independent data values.

When using a very large regularization parameter λ, the contribution of $R_j'[b]R_j'[b]^*$ can be neglected and we end up with essentially (i.e. up to the scaling factor λ^{-1}) calculating

$$\delta b = -R_j'[b]^* R_j(b). \tag{41}$$

For this update direction we have

$$\mathcal{J}_j(b + \kappa \delta b) = \mathcal{J}_j(b) - \kappa \|R_j'[b]^* R_j(b)\|_F^2 + O(\|\kappa \delta b\|_F^2) \tag{42}$$

such that it is also a descent direction for (36).

Regardless which of the update formulas (39) or (41) we use, we will need to apply the adjoint linearized residual operator $R_j'[b]^*$ to some vector in the data space. In the following we will derive and test efficient schemes for doing this (the basic propagation–backpropagation scheme), and moreover we will derive a new regularization scheme for this backpropagation technique.

A standard method for deriving *regularization schemes* is to explicitly try to minimize a cost functional which incorporates, in addition to the usual least squares data misfit, a Tikhonov–Phillips regularization term:

$$\mathcal{J}_{TP}(b) = \mathcal{J}(b) + \frac{\eta}{2}\|b\|_\alpha^2, \tag{43}$$

or, in a single step fashion

$$\mathcal{J}_{TP,j}(b) = \mathcal{J}_j(b) + \frac{\eta}{2}\|b\|_\alpha^2 \tag{44}$$

where $\eta > 0$ is the regularization parameter and $\|.\|_\alpha$ indicates some norm or semi-norm, e.g. $\|b\|_\alpha = \|\nabla b\|_{L_2}$ [Vog02]. Using this approach, the cost functional is changed significantly with the goal of obtaining in a stable way a global minimizer. We do not want to take this route, but prefer instead to keep working with the *original* least-squares cost functional (36) which only involves the *data* fit. We will minimize this cost functional by restricting the search to elements of a smaller function space, which is an alternative form of regularization.

The regularization scheme will be derived and discussed in details in Sect. 7. In the following, we will present the basic structure of our inversion method, and we will derive practical ways of applying the adjoint linearized residual operator $R_j'[b]^*$ to vectors R_j in the data space Z. This will lead us to the propagation–backpropagation technique which is considered in this paper.

6.2 Efficient Calculation of the Adjoint of the Linearized Residual Operator

In this section we want to show how the adjoint residual operator is applied in the case where many receiver positions are used simultaneously for a given source and where the medium is assumed to be possibly inhomogeneous but isotropic. A formal derivation of this result can be found in [DBBP1].

Let $\hat{C}_j$ denote some suitably chosen approximation to the operator C_j, which might depend on the source index j. For the inversion, we will need some efficient way to compute

$$R_j'(b)^* \zeta_j$$

for a given data vector $\zeta_j = \hat{C}_j R_j(b) \in Z$. Let

$$\zeta_j = (\mathbf{h}_{j,1}, \ldots, \mathbf{h}_{j,N}) \in Z \tag{45}$$

be the data vector, which in the given step j of the inversion scheme usually consists of the (filtered) residual values of all receiver positions which correspond to the source $\mathbf{q}_j$. We will in the following omit all dimensional factors (for example $\gamma_{\mathbf{H}}$) to simplify the notation. Then $R'_j(b)^*\zeta_j \in F$ is given by

$$[R'_j(b)^*\zeta_j](\mathbf{x}) = \overline{\mathbf{E}_j(\mathbf{x})} \cdot \mathcal{E}_j(\mathbf{x}), \tag{46}$$

where $\mathbf{E}_j(\mathbf{x})$ is the solution of the direct problem

$$\Lambda_M(b) \begin{pmatrix} \mathbf{E}_j \\ \mathbf{H}_j \end{pmatrix} = \mathbf{q}_j, \tag{47}$$

and $\mathcal{E}_j(\mathbf{x})$ solves the corresponding adjoint problem

$$\Lambda_M^*(b) \begin{pmatrix} \mathcal{E}_j \\ \mathcal{H}_j \end{pmatrix} = \begin{pmatrix} 0 \\ \sum_{n=1}^N \mathbf{h}_{j,n}\delta(\mathbf{x} - d_n) \end{pmatrix}, \tag{48}$$

with

$$\Lambda_M(b) = \begin{pmatrix} -b & \nabla\times \\ \nabla\times & a \end{pmatrix}, \quad \Lambda_M^*(b) = \begin{pmatrix} -\bar{b} & \nabla\times \\ \nabla\times & \bar{a} \end{pmatrix}, \tag{49}$$

$$\begin{aligned} b &= i\omega\epsilon + \sigma, & \bar{b} &= -i\omega\epsilon + \sigma, \\ a &= i\omega\mu, & \bar{a} &= -i\omega\mu. \end{aligned} \tag{50}$$

We see that we only have to solve one forward problem (47) and one adjoint problem (48) in order to calculate $R'_j(b)^*\zeta_j$ in an efficient way (compare Fig. 2).

6.3 Iterative Algorithm for the Adjoint Field Inversion

We will now describe in detail the algorithm which we use in the numerical experiments. In brief algorithmic form, the iterative scheme can be written as

$b_p^{(0)} = b^{(0)}$

sweep_loop: DO $i = 1, I_{max}$

$\qquad b_0^{(i)} = b_p^{(i-1)}$

$\qquad$ source_loop: DO $j = 1, p$

$\qquad\qquad \delta b_j^{(i)} = -R'_j(b_{j-1}^{(i)})^*\hat{C}_j R_j(b_{j-1}^{(i)})$

$\qquad\qquad b_j^{(i)} = b_{j-1}^{(i)} + \kappa_j \delta b_j^{(i)}$

$\qquad$ END DO source_loop

END DO sweep_loop

Here, $b^{(0)}$ is some initial guess for b and I_{max} is the total number of sweeps. The parameter κ_j is the step-size in step j, and $\hat{C}_j$ the chosen approximation to C_j. For example, $b^{(0)}$ can be taken to be the correct layered background parameter distribution. The aim will then be to detect some unknown inclusions which are imbedded in this layered background medium, see for example [DBBP1]. Notice that in one single step of the algorithm we have to solve one forward problem (47) to compute $R_j(b^{(i)}_{j-1})$, and one adjoint problem (48) to compute $\delta b^{(i)}_j$ from a given $R_j(b^{(i)}_{j-1})$.

We mention that, in the algorithm presented so far, no explicit regularization has been applied. If there is a sufficient amount of independent data available, this algorithm already yields good reconstructions. However, for practical applications it would be desirable to have an efficient toolbox of regularization schemes available which can be used for stabilizing the inversion in situations with only very few data, and in order to incorporate prior information into the reconstructions. In the following section, we propose such an efficient and flexible regularization tool. It has been tested and evaluated numerically with great success in a different but related geophysical application in [GKMD]. The numerical evaluation of this scheme in the situation of 3D EMIT still needs to be done, and is planned as part of our future research.

7 Regularization and Smoothing with Function Spaces

We have presented above the basic algorithm which recovers L_2 functions of parameters from given data such that the misfit in the data is minimized. This procedure does not incorporate explicit regularization (except of the stabilizing procedure incorporated in the operator $\hat{C}$). In some situations, it might be necessary or desirable to restrict the search for parameter functions to a smaller subset of L_2, for example of smoothly varying functions. This might be so in order to regularize the reconstruction algorithm, or in order to take into account some prior information or assumptions on the solution we are looking for. For example, the field engineer might know or assume that the parameter distribution in some region is fairly smoothly varying. Or, he might only have very few data available for the inversion, so that he wants to select a smoothly varying profile as a regularized form of the reconstructed parameter distribution. This can be easily done in our framework.

Instead of looking for parameter distributions in $L_2(\Omega)$, let us assume now that we require the parameter to be an element of the smaller subspace

$$H_1(\Omega) := \{ m \in L_2(\Omega), \ \partial_i m \in L_2(\Omega) \text{ for } i = 1, 2, 3 \}.$$

This Sobolev space is usually equipped with the standard norm

$$\|m\|_{1,1} := \left(\|m\|^2_{L_2} + \|\nabla m\|^2_{L_2} \right)^{1/2}$$

and the standard inner product

$$\langle m_1, m_2 \rangle_{1,1} := \langle m_1, m_2 \rangle_{L_2} + \langle \nabla m_1, \nabla m_2 \rangle_{L_2}.$$

For reasons explained below, we will instead prefer to work with the equivalent norm

$$\|m\|_{\alpha,\beta} := \left(\alpha \|m\|_{L_2}^2 + \beta \|\nabla m\|_{L_2}^2 \right)^{1/2}, \quad \alpha, \beta > 0$$

and its associated inner product

$$\langle m_1, m_2 \rangle_{\alpha,\beta} := \alpha \langle m_1, m_2 \rangle_{L_2} + \beta \langle \nabla m_1, \nabla m_2 \rangle_{L_2}.$$

A proper choice of the weighting parameters α and β will allow us to steer the regularization properties of our algorithm in an efficient and predictable way.

Let us denote the new parameter space $H_1(\Omega)$, when equipped with the weighted norm $\| \cdot \|_{\alpha,\beta}$, by $\hat{F}$. When using this modified space in our algorithm, we also have to adjust the operators acting on it, in particular the adjoint of the linearized residual operator. This operator is now required to map from the data space Z into $\hat{F}$. Moreover, the minimum norm solution of (38) is now taken with respect to the weighted norm $\| \cdot \|_{\alpha,\beta}$, which clearly gives us a different candidate. The necessary adjustments for our algorithm can be done as follows.

Denote as before by $R_j'[b]^*\zeta$ the image of $\zeta \in Z$ under application of the adjoint linearized residual operator as calculated in Sect. 6.2, considered as an operator mapping from Z into F. Denote furthermore by $R_j'[b]^\circ \zeta$ its image under the adjoint linearized residual operator with respect to the newly defined weighted inner product, mapping into the smaller space $\hat{F}$. With a straightforward calculation, using the definitions of the two adjoint operators

$$\langle R_j'[b]x, \zeta \rangle_Z = \langle x, R_j'[b]^*\zeta \rangle_F = \langle x, R_j'[b]^\circ \zeta \rangle_{\hat{F}}, \tag{51}$$

it follows that

$$R_j'[b]^\circ \zeta = (\alpha I - \beta \Delta)^{-1} R_j'[b]^*\zeta, \tag{52}$$

where we supplement the inverted differential operator $(\alpha I - \beta \Delta)^{-1}$ by the boundary condition $\nabla (R_j'[b]^\circ \zeta) \cdot n = 0$ on $\partial \Omega$. The symbol I stands for the identity, and Δ stands for the Laplacian operator. Equation (52) can be easily derived by applying Green's formula to the right hand side equality in (51).

In practice, the ratio $\gamma = \beta/\alpha$ (which can be considered being a 'regularization parameter') is an indicator for the 'smoothing properties' of our scheme. The larger this ratio, the more weight is put on minimizing the derivatives of our solution. Therefore, by properly choosing this ratio, we can steer the smoothness properties of our final reconstruction to a certain degree. In our numerical experiments, we will choose this ratio once, when starting the algorithm, and keep it fixed during the iterations. The other free parameter, say α, will be chosen in each individual step to scale the update properly. In our numerical experiments, we choose α such that

$$\|R_j'[b]^\circ \zeta\|_{L_2} = \|R_j'[b]^*\zeta\|_{L_2}$$

is satisfied for the current update. This possibility of scaling the updates is the main reason for keeping the parameter α throughout the calculations instead of simply putting it to 1 right at the beginning. When testing and comparing the performance of different regularization parameters γ it is practically useful that the order of magnitude of the calculated values of $R'_j[b]^\circ\zeta$ does not depend too much on γ.

Notice also that the new search directions using this modified adjoint operator are still *descent directions* for the least squares cost functional (36), as can be verified easily by replacing F by $\hat{F}$ in (40) and (42).

Practically, the scheme is implemented as follows:

γ is fixed regularization parameter

Define $\Psi = R'_j[b]^*\zeta$.

Solve $(I - \gamma\Delta)\varphi = \Psi$, $\nabla\varphi\cdot n = 0$ on $\partial\Omega$.

Define $\Phi = \frac{\varphi}{\alpha}$ with $\alpha = \frac{\|\varphi\|}{\|\Psi\|}$ (such that $\|\Phi\| = \|\Psi\|$)

Then we have $(\alpha I - \beta\Delta)\Phi = \Psi$, $\nabla\Phi\cdot n = 0$ on $\partial\Omega$,   with $\beta = \alpha\gamma$.

Put $R'_j[b]^\circ\zeta = \Phi$.   (53)

We mention that applying this regularization scheme amounts to applying the postprocessing operator $(\alpha I - \beta\Delta)^{-1}$ to the updates calculated in the previous 'unregularized' scheme. Therefore, the effect of the regularization is similar to filtering the updates with a carefully designed (iteration-dependent) filtering operator.

In the following, we want to give an interesting *additional interpretation of this regularization scheme*.

Define the cost functional

$$\hat{\mathcal{J}}(\Phi) = \frac{\kappa_1}{2}\|\Phi\|^2_{L_2} + \frac{\kappa_2}{2}\|\nabla\Phi\|^2_{L_2} + \frac{\kappa_3}{2}\|\Phi - \Psi\|^2_{L_2} \qquad (54)$$

with $\Psi = R'_j[b]^*\zeta$. Here, the third term penalizes the misfit between the unregularized update direction $\delta b = R'_j[b]^*\zeta$ and the new candidate Φ, whereas the first two terms penalize roughness of Φ. The gradient direction for this cost functional is $[(\kappa_1 + \kappa_3)I - \kappa_2\Delta]\Phi - \kappa_3\Psi$ (where the Laplace operator is again understood to be accompanied by the boundary condition $\nabla\Phi\cdot n = 0$ on $\partial\Omega$). Therefore, a necessary condition for the minimum can be stated as

$$[(\kappa_1 + \kappa_3)I - \kappa_2\Delta]\Phi = \kappa_3\Psi. \qquad (55)$$

Choosing $\kappa_3 = 1$, $\kappa_2 = \beta \geq 0$ and $\kappa_1 = \alpha - 1 \geq 0$ this amounts to calculating

$$\Phi = (\alpha I - \beta\Delta)^{-1}\Psi, \qquad (56)$$

which is equivalent to (52). Therefore, applying function space regularization as described above can be interpreted as minimizing the cost functional (54) with specifically chosen parameters κ_1, κ_2 and κ_3.

As already mentioned, this regularization scheme has been applied with great success to a different but similar inverse problem of reservoir characterization in [GKMD].

8 Numerical Examples for Electromagnetic Sensitivity Functions

8.1 Some Numerically Calculated Sensitivity Functions

In the numerical simulations presented here, we use a $40 \times 60 \times 55$ staggered grid for discretizing the physical domain of interest (see [CBB01, DBBP1] for details). The computational grid is surrounded by 10 additional cells of PML (perfectly matched layer) on each side. The total grid (including PML) which is used in our computations is therefore $60 \times 80 \times 75$ grid cells. Each of the individual grid cells in the physical domain has a size of $3.5 \times 3.5 \times 3.5\,\mathrm{m}^3$, whereas those in the PML have a size of 10 m in the direction perpendicular to the boundary, and 3.5 m in the remaining directions. The total interior domain which is modelled by the computational grid without the PML has therefore the physical size $140 \times 210 \times 192.5\,\mathrm{m}^3$, and including the PML it has the physical size $340 \times 410 \times 392.5\,\mathrm{m}^3$.

The sources and adjoint sources (receivers) are in all examples y or z-directed magnetic dipole sources of the strength $1\,\mathrm{A\,m}^2$ and with frequency $f = 1.0\,\mathrm{kHz}$. They are located in two boreholes which are a distance of 30 grid cells or 105 m apart from each other. We will only consider homogeneous media in this paper which have constant parameter distributions everywhere. The more general situation of layered media with an air/soil interface and with objects imbedded are discussed in [DBBP1, DBBP2]. Nevertheless, we will introduce a (fictitious) air/soil interface which separates two parts of our homogeneous medium, an upper part (called 'air') and a lower part (called 'soil'). The separating interface is called the 'free surface'. In our sensitivity analysis we use two different depths for the source and receiver locations, one being at about 80 m, and the other one at about 7 m below the free surface.

Figures 3–9 show some examples for numerically calculated sensitivity functions. In all situations, the parameter σ in the medium is $0.1\,\mathrm{S\,m}^{-1}$ everywhere, and the relative permittivity ϵ_r, as well as the permeability parameter μ_r, are 1.0. The source is a z-directed magnetic dipole transmitter which is positioned in one borehole at a depth of 80 m below the free surface. A second borehole containing a receiver antenna is located at the same x-coordinate, but it has a horizontal distance in the y-coordinate to the first borehole of 105 m. The z-axis points 'downward' into the earth and indicates 'depth'. The receiving antenna in the second borehole is located either at the same depth of 80 m below the free surface (Figs. 3–7), or at a depth of 7 m below the free surface (Figs. 8–9). It measures either the z-component or the y-component of the magnetic field, as it is indicated in the individual figures. The frequency is, as in all of our numerical experiments, $f = 1\,\mathrm{KHz}$. All sensitivity functions displayed here are 'isotropic sensitivity functions'. For more details regarding the behaviour of 'anisotropic sensitivity functions', we refer the reader to [DBBP2].

Displayed in Fig. 3 is the real part of the numerically calculated sensitivity function $S_z(1, z; \mathbf{y})$, which we will denote in the following simply by $S(z, z; \mathbf{y})$. (Similarly, we will denote $S_z(1, z; \mathbf{y})$ simply by $S(z, y; \mathbf{y})$, ignoring the 1 in the

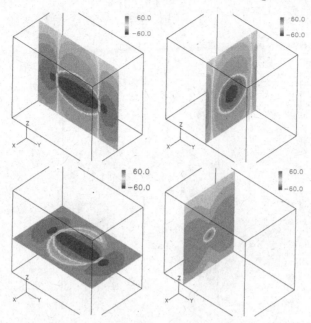

Fig. 3. Shown are different cross-sections through the *real part* of the numerically calculated isotropic sensitivity function $S(z, z; \mathbf{y})$ as a function of $\mathbf{y}$ for an infinitely extended homogeneous medium. The source and the receiver are z-directed magnetic dipoles at depth 80 m. The scaling factor is Rmax $= 3.62 \times 10^{-13}$

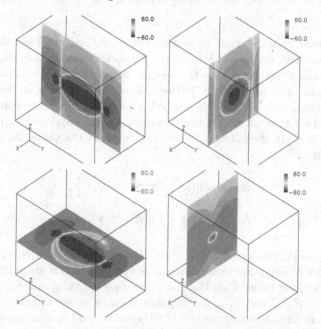

Fig. 4. Shown are the same cross-sections through the *real part* of the isotropic sensitivity function $S(z, z; \mathbf{y})$ as in Fig. 3, but now calculated analytically using the expressions derived in [DBBP2]. The scaling factor is again Rmax $= 3.62 \times 10^{-13}$. The agreement with Fig. 3 is very good

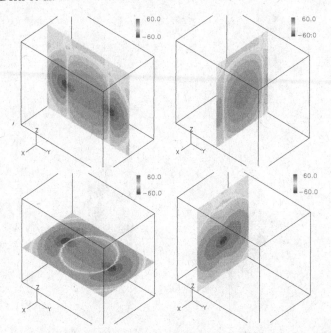

Fig. 5. Shown are different cross-sections through the *imaginary part* of the numerically calculated isotropic sensitivity function $S(z, z; \mathbf{y})$ as a function of $\mathbf{y}$ for an infinitely extended homogeneous medium. The source and the receiver are z-directed magnetic dipoles at depth 80 m. The scaling factor is Imax $= 2.96 \times 10^{-12}$

notation which indicates that we consider only one receiver.) In this simplified notation, the first argument z indicates the orientation of the dipole source, and the second argument z says that we are measuring the z-component of the magnetic field. The argument $\mathbf{y}$ is the independent variable of the sensitivity function ranging over all possible scattering positions in the medium.

We mention that all sensitivity functions in this paper are displayed on a logarithmic scale in the following sense. Let us define the values R_{max} and I_{max} according to

$$R_{max} = \max_{\mathbf{y} \in \Omega} \left\{ \left| \mathrm{Re}(S(z, z; \mathbf{y})) \right| \right\},$$

$$I_{max} = \max_{\mathbf{y} \in \Omega} \left\{ \left| \mathrm{Im}(S(z, z; \mathbf{y})) \right| \right\},$$

and analogous expressions for $S(z, y; \mathbf{y})$. Here, the maximum is taken over the finite number of computer evaluations of $S(z, z; \mathbf{y})$ which is given by the FDFD forward modelling code. This has the effect that all values of $S(z, z; \mathbf{y})$ which contribute to R_{max} and I_{max} are actually finite, whereas the theoretically calculated values would become singular at the source and receiver positions. With these reference values R_{max} and I_{max}, we define the logarithmic sensitivity functions by

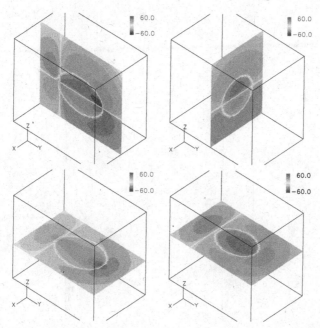

Fig. 6. Shown are different cross-sections through the *real part* of the numerically calculated isotropic sensitivity function $S(z, y; \mathbf{y})$ as a function of $\mathbf{y}$ for an infinitely extended homogeneous medium. The source is a z-directed magnetic dipole at depth 80 m, and the receiver is a y-directed magnetic dipole at depth 80 m. The scaling factor is Rmax $= 3.36 \times 10^{-13}$

$$\text{logreal}(z, z; \mathbf{y}) = -\text{sign}(\text{Re}(S(z, z; \mathbf{y}))) \left(60 - 10 \times \log \left(\frac{R_{max}}{|\text{Re}(S(z, z; \mathbf{y}))|} \right) \right)_+$$

$$\text{logimag}(z, z; \mathbf{y}) = -\text{sign}(\text{Im}(S(z, z; \mathbf{y}))) \left(60 - 10 \times \log \left(\frac{I_{max}}{|\text{Im}(S(z, z; \mathbf{y}))|} \right) \right)_+ ,$$

and analogous expressions hold for $S(z, y; \mathbf{y})$. Here, sign(x) gives the sign ('plus' or 'minus') of the real number x, the symbol $(x)_+ = \max\{0, x\}$ acts as the identity for positive values of x and delivers zero for negative values. 'Re' means 'real part', and 'Im' means 'imaginary part'. It is these logarithmic sensitivity functions which are displayed in Figs. 3–9.

In Fig. 4 we have displayed the same sensitivity functions as in Fig. 3, but now being calculated with an independent analytical expressions which has been derived in [DBBP2]. From the agreement of the two Figs. 3 and 4 we can conclude that our numerically calculated sensitivity functions are indeed correct. Additional numerical verifications have been done in [DBBP2]. Notice also that all our sensitivity functions have been evaluated on the grid positions of our staggered FDFD grid.

Figure 5 displays the *imaginary part* of the sensitivity function for the same setup as in Figs. 3 and 4. In Figs. 6 and 7 the real and imaginary part of the

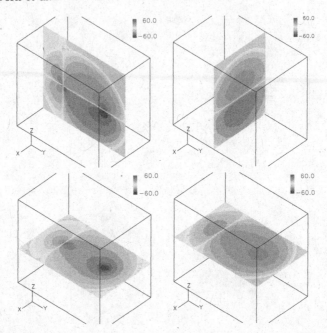

Fig. 7. Shown are different cross-sections through the *imaginary part* of the numerically calculated isotropic sensitivity function $S(z, y; \mathbf{y})$ as a function of $\mathbf{y}$ for an infinitely extended homogeneous medium. The source is a z-directed magnetic dipole at depth 80 m, and the receiver is a y-directed magnetic dipole at depth 80 m. The scaling factor is Imax $= 2.96 \times 10^{-12}$

sensitivity functions are displayed, respectively, which correspond to the situation where the transmitter is a z-directed magnetic dipole, but the receiver is a y-directed magnetic dipole. Finally, Figs. 8 and 9 consider the situation where both the transmitter and the receiver are z-directed magnetic dipoles, but the transmitter is located at a depth of 80 m whereas the receiving antenna is located in the second borehole at a depth of only 7 m.

Certainly, the idealized homogeneous sensitivity functions as presented here only can give some indication about the true sensitivity structure in a more realistic scenario where also the air–soil interface is taken into account, and where a layered medium is assumed for the soil. Such a scenario has been investigated in more details in [DBBP2].

8.2 A Short Discussion of Sensitivity Functions

We have presented in Sect. 8.1 some results of our numerical experiments for calculating sensitivity functions. In the following, we want to indicate a few problems which come up in typical imaging scenarios, and where the study of sensitivity functions might be very useful in order to address these problems successfully.

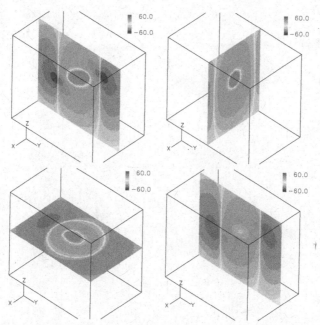

Fig. 8. Shown are different cross-sections through the *real part* of the numerically calculated isotropic sensitivity function $S(z, y; \mathbf{y})$ as a function of $\mathbf{y}$ for an infinitely extended homogeneous medium. The source is a z-directed magnetic dipole at depth 80 m, and the receiver is a z-directed magnetic dipole at depth 7 m. The scaling factor is Rmax = 4.00×10^{-13}

First, when imaging a geophysical test site, some decision has to be made about which degree of anisotropy should be expected in the given situation. For a discussion of some implications of electrical anisotropy in applications see also Weidelt [Wei95]. It can be observed in our numerical experiments (not shown here, but see [DBBP2]) that the shapes of the x, y and z-component sensitivity functions typically differ from each other, and certain source-receiver constellations are very sensitive to some of these components and much less sensitive to others. This indicates that it should in general be possible to resolve anisotropic structures in the earth from electromagnetic data, provided that a sufficient amount of appropriately gathered data is available. On the other hand, it indicates that the treatment of strongly anisotropic media with an isotropic imaging approach can lead to inaccurate reconstructions because of the related misinterpretations of the data.

Second, in basically all imaging problems the question comes up which degree of resolution can be expected from the reconstructions. In our situation, since the reconstructions are essentially results of a series of backprojection steps, only features of the medium can be recovered which are actually represented by the sensitivity functions. This gives us some information about the resolution which we can expect. If we want to further quantify the degree of resolution, we could use for example the concept of a 'resolution matrix' which has been discussed

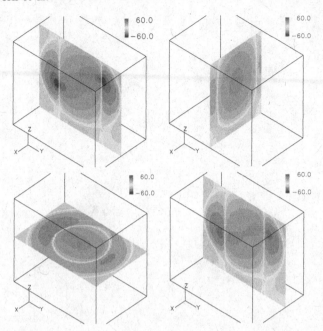

Fig. 9. Shown are different cross-sections through the *imaginary part* of the numerically calculated isotropic sensitivity function $S(z, y; \mathbf{y})$ as a function of $\mathbf{y}$ for an infinitely extended homogeneous medium. The source is a z-directed magnetic dipole at depth 80 m, and the receiver is a z-directed magnetic dipole at depth 7 m. The scaling factor is Imax $= 1.22 \times 10^{-12}$

in the similar situation of hydrology by Vasco et al. [VDL97]. The sensitivity functions play a natural role in this approach, which is not very surprising since we would intuitively expect that for example the broadening of the sensitivity functions some distance away from the sources and receivers will give rise to a decrease in the resolution of our imaging method at these points. A similar observation was already mentioned above in the discussion of the broadening of the sensitivity functions in DOT which leads to a decreased resolution compared to X-ray tomography. See for example [Ar95a, Ar95b, Ar99, Dor00].

Third, the knowledge of the general form of the sensitivity functions can be used in order to design experiments more efficiently. See for example Maurer et al. [MBC00] and the references therein for a discussion of possible design strategies for electromagnetic geophysical surveys. The Fréchet derivative plays an important role in these strategies, and we have seen above that the Fréchet derivative is directly related to our sensitivity functions. A possible design strategy would for example be to combine sources and receivers in a way such that the resulting sensitivity structure is focused on certain regions of interest in the earth and is small in other areas. This typically has to be done under certain constraints on the availability and number of boreholes, on the given source pattern and receiver characteristics, and other test site specific constraints. We refer for more details again to [MBC00] and the references given there.

Acknowledgments

Part of the work presented here has been performed during the research visit of O.D. at UCLA, Los Angeles, in the fall of 2003 for the IPAM special program on 'imaging'. The support of IPAM is gratefully acknowledged.

References

[AM95a]   Alumbaugh D L and Morrison H F 1995 Monitoring subsurface changes over time with cross-well electromagnetic tomography *Geophysical Prospecting* **43** 873–902

[AM95b]   Alumbaugh D L and Morrison H F 1995 Theoretical and practical considerations for crosswell electromagnetic tomography assuming cylindrical geometry *Geophysics* **60** 846–870

[AB01]   Ammari H and Bao G 2001 Analysis of the scattering map of a linearized inverse medium problem for electromagnetic waves *Inverse Problems* **17** 219–234

[Ar95a]   Arridge S R 1995 Photon-measurement density functions. Part I: Analytical forms *Applied Optics* **34** (34) 7395-7409

[Ar95b]   Arridge S R and Schweiger M 1995 Photon-measurement density functions. Part II: Finite-element-method calculations *Applied Optics* **34** (31) 8026-8037

[Ar99]   Arridge S (1999) Optical Tomography in Medical Imaging *Inverse Problems* **15** R41-R93

[AP98]   Ascher U M and Petzold L R (1998) *Computer Methods for Ordinary Differential Equations and Differential-Algebraic Equations* (SIAM: Philadelphia)

[BK90]   Berryman, J. G., and R. V. Kohn, 1990, Variational constraints for electrical impedance tomography, *Phys. Rev. Lett.* **65**, 325–328

[BBP96]   Borcea L, Berryman J G, and Papanicolaou G C 1996 High Contrast Impedance Tomography, *Inverse Problems* **12** 935–958

[BBP99]   Borcea L, Berryman J G, and Papanicolaou G C 1999 Matching pursuit for imaging high contrast conductivity *Inverse Problems* **15** 811–849

[Bor02]   L. Borcea 2002 Electrical Impedance Tomography, *Inverse Problems* **18** R99–R136

[BB99]   Buettner H M and Berryman J G 1999 An electromagnetic induction tomography field experiment at Lost Hills, CA, in the *Proceedings of SAGEEP*, Oakland, CA, March 14–18, 663–672

[CBB01]   N. J. Champagne II, J. G. Berryman, and H. M Buettner, FDFD (2001) A 3D finite-difference frequency-domain code for electromagnetic induction tomography *J. Comput. Phys.* **170** 830–848

[Cla76]   Claerbout, J. F., 1976, *Fundamentals of Geophysical Data Processing: With Applications to Petroleum Prospecting*, McGraw-Hill, New York.

[DRLN]   Daily, W. D., A. Ramirez, D. LaBrecque, and J. Nitao, 1992, Electrical resistivity tomography of vadose water movement, *Water Resources Res.* **28**, 1429–1442

[DA89]   Davis, J. L., and A. P. Annan, 1989, Ground penetrating radar for high resolution mapping of soil and rock stratigraphy, *Geophysical Prospect.* **37**, 531–551

[DS96]   Dennis J E and Schnabel R B 1996 *Numerical Methods for Unconstrained Optimization and Nonlinear Equations,* (Reprint in the SIAM Classics in Applied Mathematics Series No. 16)

[DDNPS]   Dierkes T, Dorn O, Natterer F, Palamodov V and Sielschott H 2002 Fréchet derivatives for some bilinear inverse problems, *SIAM J. Appl. Math.* **62** (6) 2092-2113

[DL79]   Dines K A and Lytle R J 1979 Computerized geophysical tomography *Proc. IEEE* **67** 1065-73

[Dor98]   Dorn O 1998 A transport-backtransport method for optical tomography 1998 *Inverse Problems* **14** 1107-1130

[DBBP1]   Dorn O, Bertete-Aguirre H, Berryman J G and Papanicolaou G C 1999 A nonlinear inversion method for 3D-electromagnetic imaging using adjoint fields *Inverse Problems* **15** 1523-1558

[Dor00]   Dorn O 2000 Scattering and absorption transport sensitivity functions for optical tomography *Optics Express* **7** (13) 492–506

[DBBP2]   Dorn O, Bertete-Aguirre H, Berryman J G and Papanicolaou G C 2002 Sensitivity analysis of a nonlinear inversion method for 3D electromagnetic imaging in anisotropic media *Inverse Problems* **18** 285-317

[FO96]   Farquharson C G and Oldenburg D W 1996 Approximate sensitivities for the electromagnetic inverse problem *Geophysical Journal International* **126** 235-252

[FMA92]   Fisher, E., G. A. McMechan, and A. P. Annan, 1992, Acquisition and processing of wide-aperature ground penetrating radar data, *Geophysics* **57**, 495–504.

[GKMD]   Gonzáles-Rodríguez P, Kindelan M, Moscoso M and Dorn O 2005 History matching problem in reservoir engineering using the propagation-backpropagation method *Inverse Problems* **21**, 565–590

[HGS93]   Habashy T M, Groom R W and Spies B R 1993 Beyond the Born and Rytov approximations: A nonlinear approach to electromagnetic scattering *J. Geophys. Res.* **98** 1759–1775

[HAO00]   Haber E, Ascher U and Oldenburg D 2000 On optimization techniques for solving nonlinear inverse problems *Inverse Problems* **16** 1263-1280

[JW57]   Jenkins F A and White H E 1957 *Fundamentals of Optics* (McGraw-Hill: New York) p 300

[MBC00]   Maurer H, Boerner D E and Curtis A 2000 Design strategies for electromagnetic geophysical surveys *Inverse Problems* **16** 1097-1117

[MO90]   McGillivray P R and Oldenburg D W 1990 Methods for calculating Frechet derivatives and sensitivities for the non-linear inverse problem: a comparative study *Geophysical Prospecting* **38** 499-524

[MOEH]   McGillivray P R, Oldenburg D W, Ellis R G, and Habashy T M 1994 Calculation of sensitivities for the frequency-domain electromagnetic problem *Geophysical Journal International* **116** 1-4

[Nat86]   Natterer F 1986 *The Mathematics of Computerized Tomography* (Stuttgart: Teubner)

[NW95]   Natterer F and Wübbeling F 1995 A propagation-backpropagation method for ultrasound tomography *Inverse Problems* **11** 1225-1232

[Nat96]   Natterer F (1996) Numerical Solution of Bilinear Inverse Problems *Preprints* "Angewandte Mathematik und Informatik" 19/96-N Münster

[NW01]   Natterer F and Wübbeling F *Mathematical Methods in Image Reconstruction,* Monographs on Mathematical Modeling and Computation 5, SIAM 2001

[NA97]   Newman G A and Alumbaugh D L 1997 Three-dimensional massively par-
           allel electromagnetic inversion I. Inversion *Geophys. J. Int.* **128** 345–354

[NW99]   Nocedal J and Wright S J 1999 *Numerical Optimization*, (Springer: New
           York)

[RDLOC]   Ramirez A, Daily W, LaBrecque D, Owen E and Chesnut D 1993 Moni-
           toring an underground steam injection process using electrical resistance
           tomography *Water Resources Res.* **29** 73–87

[SH95]   Spies B P and Habashy T M 1995 Sensitivity analysis of crosswell electro-
           magnetics *Geophysics* **60** 834–845

[VDL97]   Vasco D W, Datta-Gupta A and Long J C S 1997 Resolution and uncer-
           tainty in hydrologic characterization *Water Resources Res.* **33** 379–397

[Vog02]   Vogel C R *Computational Methods for Inverse Problems* (SIAM series
           'Frontiers in Applied Mathematics' 2002)

[Wei95]   Weidelt P 1995 Three-dimensional conductivity models: implications of
           electrical anisotropy. In Oristaglio M and Spies B (ed) *International
           Sumposium on Three-dimensional Electromagnetics, Oct 4-6, 1995, at
           Schlumberger-Doll Research, Ridgefield/Connecticut, USA*, pp 1-11

[WAMBL]   Wilt, M. J., D. L. Alumbaugh, H. F. Morrison, A. Becker, K. H. Lee,
           and M. Deszcz-Pan, 1995a, Crosswell electromagnetic tomography: System
           design considerations and field results, *Geophysics* **60**, 871–885

[WMBTL]   Wilt, M. J., H. F. Morrison, A. Becker, H.-W. Tseng, K. H. Lee, C.
           Torres-Verdin, and D. Alumbaugh, 1995b, Crosshole electromagnetic to-
           mography: A new technology for oil field characterization, *The Leading
           Edge* **14**, March, 1995, pp. 173–177

[ZTB96]   Zhdanov, M. S., P. Traynin, and J. R. Booker, 1996, Underground imaging
           by frequency-domain electromagnetic migration, *Geophysics* **61**, 666–682

Polarization-Based Optical Imaging

Miguel Moscoso

Gregorio Millán Institute of Fluid Dynamics, Nanoscience and Industrial
Mathematics, Universidad Carlos III de Madrid, Avda. de la Universidad 30,
28911 Leganés, Madrid, Spain
moscoso@math.uc3m.es

Summary. In this study I present polarization effects resulting from the reflection
and transmission of a narrow beam of light through biological tissues. This is done
numerically with a Monte Carlo method based on a transport equation which takes into
account polarization of light. We will show both time-independent and time-dependent
computations, and we will discuss how polarization can be used in order to obtain
better images. We consider biological tissues that can be modeled as continuous media
varying randomly in space, containing inhomogeneities with no sharp boundaries.

1 Introduction

Optical imaging is proving to be a potentially useful non-invasive technique for
the detection of objects in the human body. The ultimate goal is the detection of
millimeter-sized cancerous tissue before it spreads into the surrounding healthy
tissue [1–3]. Cancerous tissue absorbs more light than healthy tissue due to
its higher blood content, giving rise to a perturbation in the measured light
intensity. The main problem in using light as a diagnostic tool is that it is strongly
scattered by normal tissue, resulting in image blurring. Various techniques have
been devised to overcome this difficulty, e.g., time gating, optical coherence,
confocal detection, etc. All these techniques distinguish between weakly and
strongly scattered photons, i.e., those scattered through small angles and those
scattered through large angles.

Recently, there has been a considerable interest in the polarization properties
of the reflected and transmitted light. Multiple scattering gives rise to diffusion
and depolarization. The effectiveness of polarization-sensitive techniques to discriminate between weakly and strongly scattered photons has been demonstrated
experimentally by Schmitt et al. [4], Rowe et al. [5], and Demos et al. [6], among
others. We have reported numerical evidence for this previously using the theory
of radiative transfer taking into account polarization of light [7].

The theory of radiative transfer, employing Stokes parameters, can be used
to describe light propagation through a medium. Analytical solutions are not
known except in simple particular cases, such as plane parallel atmospheres

with a constant net flux [8]. When the incident laser beam is narrow no solution is known. Therefore we have developed a Monte Carlo method to obtain the spatial distribution of the total intensity and of the polarization components of the transmitted and reflected beams. Other methods for solving the vector radiative transfer equation have been explored in [9].

In [9] we presented a complete discussion of Chebyshev spectral methods for solving radiative transfer problems. In this method, we approximated the spatial dependence of the intensity by an expansion of Chebyshev polynomials. This yields a coupled system of integro-differential equations for the expansion coefficients that depend on angle and time. Next, we approximated the integral operation on the angle variables using a Gaussian quadrature rule resulting in a coupled system of differential equations with respect to time. Using a second order finite-difference approximation, we discretized the time variable. We solved the resultant system of equations with an efficient algorithm that makes Chebyshev spectral methods competitive with other methods for radiative transfer equations.

Section 2 contains the theory of radiative transfer employing Stokes parameters. In Sect. 3 we describe our model where we represent a biological tissue as a medium with weak random fluctuations of the dielectric permittivity $\epsilon(\mathbf{r}) = \epsilon_0[1 + \delta\epsilon(\mathbf{r})]$, where $\mathbf{r}$ denotes position and $\delta\epsilon(\mathbf{r})$ is the continuous random fractional permittivity fluctuation. In Sect. 4 we describe a Monte Carlo method for three-dimensional problems which take into account the vector nature of light. Section 5 describes our numerical calculations. Section 6 contains our conclusions.

2 Radiative Transfer Equation

The theory of radiative transport employing Stokes parameters can be used to describe how light propagates through an absorbing scattering medium. Both the absorption coefficient Σ_a and the scattering coefficient Σ_s of the background medium, can be determined from reflection and transmission measurements. Absorption of light in tissue is due to natural chromophores, such as the heme pigment of hemoglobin, the cytochrome pigments of the respiratory chain in mitochondria, and others [10]. Scattering in soft tissue is due to random low contrast inhomogeneities of the refractive index, of the order of 5% [11]. Due to interaction with the inhomogeneities, light waves with wave vector $\mathbf{k}'$ at point $\mathbf{r}$ can scatter into any direction $\hat{\mathbf{k}}$ with wave vector $\mathbf{k}$, where $\hat{\mathbf{k}} = \mathbf{k}/|\mathbf{k}|$. The statistical properties of the light propagating through the medium can be specified by the Stokes vector

$$\mathbf{I} = \begin{pmatrix} I \\ Q \\ U \\ V \end{pmatrix}, \tag{1}$$

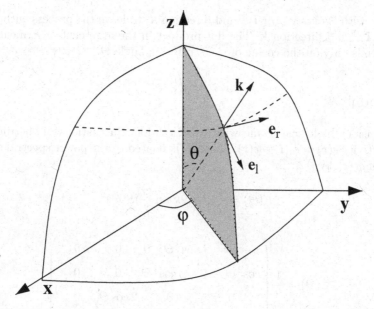

Fig. 1. Coordinate system for the Stokes vector. The *shaded plane* $\phi = contant$ containing the wavevector **k** is called the meridian plane. We choose $\mathbf{e}_l$ to lie in this plane and to be orthogonal to **k**

where

$$
\begin{aligned}
I &= \langle E_l E_l^* + E_r E_r^* \rangle, \\
Q &= \langle E_l E_l^* - E_r E_r^* \rangle, \\
U &= \langle E_l E_r^* + E_r E_l^* \rangle, \\
V &= i \langle E_l E_r^* - E_r E_l^* \rangle.
\end{aligned} \tag{2}
$$

Here E_l and E_r are the complex amplitudes of the electric field referred to an orthonormal system $(\hat{\mathbf{k}}, \mathbf{e}_l, \mathbf{e}_r)$; see Fig. 1. I is the total intensity. The Stokes parameter Q is the difference between the intensities transmitted through two linear analyzers in orthogonal directions. U is defined like Q, but with reference to linear analyzers rotated by 45° with respect to those of Q. V is the difference between the intensities passing through right and left circular analyzers.

The Stokes vector $\mathbf{I}(\mathbf{x}, \hat{\mathbf{k}}, t)$, defined for all directions $\hat{\mathbf{k}}$ at each point $\mathbf{x}$ and time t, satisfies the transport equation [8]

$$
\frac{1}{v} \frac{\partial \mathbf{I}(\mathbf{x}, \hat{\mathbf{k}}, t)}{\partial t} + \hat{\mathbf{k}} \cdot \nabla_{\mathbf{x}} \mathbf{I}(\mathbf{x}, \hat{\mathbf{k}}, t) + (\Sigma^{(s)}(\mathbf{x}) + \Sigma^{(a)}(\mathbf{x})) \mathbf{I}(\mathbf{x}, \hat{\mathbf{k}}, t) =
$$
$$
\Sigma^{(s)}(\mathbf{x}) \int F(\hat{\mathbf{k}} \cdot \hat{\mathbf{k}}') \mathbf{I}(\mathbf{x}, \hat{\mathbf{k}}', t) d\hat{\mathbf{k}}' . \tag{3}
$$

Here v, $\Sigma^{(s)}$, and $\Sigma^{(a)}$ are the velocity of the light in the medium, the scattering coefficient, and the absorption coefficient, respectively. The 4×4 scattering matrix $F(\hat{\mathbf{k}} \cdot \hat{\mathbf{k}}')$ in (3) describes the probability for a photon entering a scattering process with Stokes vector $\mathbf{I}^{(i)}$ and direction $\hat{\mathbf{k}}'$ to leave the process with Stokes vector $\mathbf{I}^{(s)}$ and direction $\hat{\mathbf{k}}$. The dot product in the argument of F means that it depends only on the cosine of the scattering angle Θ.

3 Model

We consider biological tissue with weak random fluctuations of the dielectric permittivity $\epsilon(\mathbf{r}) = \epsilon_0[1 + \delta\epsilon(\mathbf{r})]$. Then F is related to the power spectral density $\hat{R}(q)$ of $\delta\epsilon(\mathbf{r})$ by [12]:

$$F(\Theta) = \frac{\pi}{2}k^4\hat{R}(2k\sin\frac{\Theta}{2})S(\Theta) . \tag{4}$$

Here

$$S(\Theta) = \frac{1}{2}\begin{pmatrix} 1 + \cos^2\Theta & \cos^2\Theta - 1 & 0 & 0 \\ \cos^2\Theta - 1 & 1 + \cos^2\Theta & 0 & 0 \\ 0 & 0 & 2\cos\Theta & 0 \\ 0 & 0 & 0 & 2\cos\Theta \end{pmatrix} \tag{5}$$

is the Rayleigh scattering matrix. $\hat{R}$ contains statistical information about the random inhomogeneities of the medium, which can be modeled or obtained experimentally [11].

$S(\Theta)$ in (5) is referred to the scattering plane, i.e., the plane through the incident and scattered directions. It is more convenient to express S with respect to the directions $\mathbf{e}_l$, both incident and scattered, parallel to the meridian plane as in Fig. 1. We refer to [8] for more details about the form of the scattering matrix in this basis, and to [13] and [14] for a discussion of depolarization of multiply scattered light.

Equation (4) is the basic expression for the scattering matrix in a weakly fluctuating random medium, and relates the angular scattering pattern to the statistical characteristics of the medium. The scattering pattern is the product of the matrix $S(\Theta)$, which is a purely geometrical factor, and the scalar function $\hat{R}(2k\sin\Theta/2)$, which involves the statistical properties of the medium. The power spectral density function $\hat{R}$ can be obtained experimentally [11]. However, in many practical situations the spectral characteristics of the fluctuations can be modeled by simple functions. For mathematical simplicity we consider a model with only two parameters: the relative strength of the fluctuations, ε, and the correlation length, l. We use an exponential form for the correlation function,

$$R(r) = \varepsilon^2 e^{-r/l} . \tag{6}$$

In this case the power spectral density function is given by

$$\hat{R}(k) = \frac{\varepsilon^2 l^3}{\pi^2(1 + k^2 l^2)^2} . \tag{7}$$

The exponential correlation function (6) is a especial case of the von Karman correlation function

$$R(r) = \frac{\varepsilon^2 2^{1-\kappa}}{\Gamma(\kappa)} (\frac{r}{l})^\kappa K_\kappa(\frac{r}{l}), \quad \kappa \in [0, 0.5] \qquad (8)$$

for $\kappa = 0.5$, used to model the spectral characteristics of the fluctuations in turbulent fluids. In (8), Γ is the gamma function and K_κ is the modified Bessel function of the second kind of order κ. The von Karman correlation function is commonly used to model the spectral characteristics of the fluctuations in turbulence media.

We note that the power spectral density function (7) becomes independent of k for $kl \ll 1$, so then the factor $\hat{R}$ in (4) is independent of Θ. $\hat{R}(2k \sin \Theta/2)$ is proportional to $(2kl \sin \Theta/2)^{-4}$ for $kl \gg 1$, so as kl increases, the scattering becomes more peaked in the forward direction.

The mean value of the cosine of the scattering angle, called the anisotropy factor, is defined by

$$g(k) = \frac{\pi^2 k^4}{2\Sigma_s(k)} \int_{-1}^{1} \cos\Theta \hat{R}(k\sqrt{2 - 2\cos\Theta})[1 + \cos^2\Theta] d(\cos\Theta) . \qquad (9)$$

It is one of the parameters used to describe the optical properties of a medium. For the correlation function (6) with kl equal to 0, 1, 2 or 5, g is equal to 0, 0.30, 0.48, or 0.66, respectively. We note that g is typically larger than these values in biological tissues (> 0.70 for wavelengths in the visible). There, the majority of the scattering takes place from structures within the cell, like mitochondria (diameter of the order 0.3–0.7 μm) or lysosomes and peroxisomes (diameter of the order 0.2–0.5 μm) [15, 16]. More complicated correlation functions than (6) could yield larger values of g for wavelengths in the visible.

In the next section we describe a Monte Carlo Method to solve the radiative transport equation (3).

4 Monte Carlo Method

Monte Carlo methods are based on the stochastic nature of the propagation of light in a random medium. They simulate photon histories according to fundamental probability laws which depend on the scattering medium. Their main advantage is their relative simplicity and their ability to handle complicated geometries. They also provide exact solutions (up to statistical errors) of the radiative transport equation (3). Their main drawback is that they require a large number of photons to obtain statistical accuracy for large optical depths. The rate of convergence is given by the *central limit* theorem of probability theory. It states that for N independent random variables I^i from the same distribution, with finite mean m and variance σ^2, the average $I_N = \sum_i^N I^i/N$ is normally distributed about m with variance σ^2/N. This means that I_N converges to the expected value at the rate $const./\sqrt{N}$. To make it converge faster several variance reduction techniques can be used [7].

We use two basic approaches to improve the convergence [17, 18]. First, we modify the random walk sampling to follow more particles in *important* regions where the scoring contributes more to the final tally. This is especially relevant in computing light transmission through a medium of large optical thickness. In that case, only a small fraction of particles penetrates into the deeper regions of the medium, producing bad statistics in the computation. Second, we modify the scoring method to maximize the statistical information of a given random walk. In doing this, a trajectory is no longer associated with a single particle. The replacement of the analog simulation by a non-analog one is usually called a *variance reduction technique*.

In the next subsection we describe the simplest case of the analog simulation. In Sect. 4.2 we explain the geometry splitting and Russian roulette, whose objective is to spend more time sampling particles in important regions and less time sampling particles in unimportant regions. In Sect. 4.3 we explain the point detector technique, which is a deterministic estimate of the flux at a point from each collision.

4.1 Analog Monte Carlo Sampling

Consider a non-absorbing homogeneous medium with scattering coefficient Σ_s. The history of each photon is initiated at time $t = 0$ by releasing it from the point $\mathbf{r}_{in} = (x_{in}, y_{in}, 0)$ with direction $\hat{\mathbf{k}}_{in} = (0, 0, 1)$ along the inward normal to the surface of the medium, and with a specified polarization state. When the beam is linearly polarized, it is 100% polarized parallel to the (x, z) plane, so the Stokes vector can be written as $[I_{in}, Q_{in}, U_{in}, V_{in}] = [1, 1, 0, 0]$. When the beam is circularly polarized, the Stokes vector is $[I_{in}, Q_{in}, U_{in}, V_{in}] = [1, 0, 0, 1]$. We will consider only linear initial polarization here.

The path length traveled by the photon before undergoing scattering is given by

$$l_{coll} = -ln(\xi)/\Sigma_s , \qquad (10)$$

where ξ is a random number uniformly distributed in the interval $(0, 1)$. The photon's scattering position $\mathbf{r}_1$ is

$$\mathbf{r}_f = \mathbf{r}_{in} + \hat{\mathbf{k}}_{in} l_{coll} , \qquad (11)$$

with $\hat{\mathbf{k}}_{in} = (\sin \theta_{in} \cos \phi_{in}, \sin \theta_{in} \sin \phi_{in}, \cos \theta_{in})$. Next, the scattering direction $\hat{\mathbf{k}}_f$ is determined according to the probability density function [7, 19]

$$P(\hat{\mathbf{k}}_f, \hat{\mathbf{k}}_{in}, Q_{in}, U_{in}) = \frac{1}{\Sigma_s} \left[\bar{F}_{11} + \bar{F}_{12} Q_{in} + \bar{F}_{13} U_{in} \right] . \qquad (12)$$

Here $\bar{F}_{ij}$, with $i, j = 1, \dots, 4$, represents the (i, j) element of the scattering matrix $\bar{F}$ referred to the meridian plane (plane of constant ϕ_d) as plane of reference for the Stokes vectors. See [8] for the required rotation. P depends on the direction of incidence and scattering, and on the incoming polarization state. The dependence on the polarization state is essential because linearly polarized photons cannot be scattered in the direction of oscillation of the electric field. This property, related to the vector nature of light, is lost if the probability density function does not depend on the polarization. Note that P does not depend on V_{in} because $F_{14} = 0$ in weakly fluctuating media. P has been normalized so that for all incident directions and polarization states

$$\int P(\hat{\mathbf{k}}_f, \hat{\mathbf{k}}_{in}, Q_{in}, U_{in})d\hat{\mathbf{k}}_f = 1 \,, \tag{13}$$

This equation expresses particle conservation in each scattering event. We note that in (13), the integrals of the terms multiplied by Q_{in} and U_{in} are zero.

Once the new direction $\hat{\mathbf{k}}_1$ is chosen, the outgoing Stokes vector is

$$\mathbf{I}_f = \frac{1}{\Sigma_s P(\hat{\mathbf{k}}_f, \hat{\mathbf{k}}_{in}, Q_{in}, U_{in})}\bar{F}(\hat{\mathbf{k}}_f, \hat{\mathbf{k}}_{in})\mathbf{I}_{in} \,. \tag{14}$$

Equation (14) does not depend on $\hat{R}(k)$. The normalization in (14) assures that the intensity of the scattered photon remains equal to the incident one. From (14) it also follows that if the incident photon is completely polarized then the scattered photon is also completely polarized.

Next, a new path length is calculated from (10) and the tracing continues using (11)–(14) until the photon exit the medium.

If a photon reaches the detector at an angle with the normal to the sample surface that is less than its semiaperture, then its contribution to the quantities of interest (intensity and polarization) are stored. If the incident beam is circularly polarized, two quantities are computed: the total intensity I and the circular component V. In the case of a linearly polarized incident beam, three intensities are computed: the total intensity I, the parallel-polarized component I_l, and the cross-polarized component I_r. If a photon of intensity $dI = 1$ arrives at the detector with polar direction (θ_d, ϕ_d), these two last intensities can be calculated easily from the Stokes parameters Q_d and U_d of the photon. The direction of polarization makes an angle

$$\chi_d = \pm 1/2 \arctan(U_d/Q_d) \tag{15}$$

with the meridian plane. The sign is positive for transmission, and negative for reflection. The difference in signs is due to the change of sign of U when the z component of the wave-vector $\hat{\mathbf{k}}$ changes sign. We have chosen $\mathbf{e}_l \times \mathbf{e}_r = \hat{\mathbf{k}}$ (see Fig. 1). If $\mathbf{e}_l \times \mathbf{e}_r = -\hat{\mathbf{k}}$ instead, then the signs must be changed in (15). The contribution of the photon to the transmitted intensities is:

$$I = dI \,, \tag{16}$$
$$I_l = dI[\cos^2(\phi_d + \chi_d)\cos^2\theta_d + \sin^2\theta_d] \,, \tag{17}$$
$$I_r = dI[\sin^2(\phi_d + \chi_d)\cos^2\theta_d + \sin^2\theta_d] \,. \tag{18}$$

To compute the reflected intensities, we must change χ_d to $-\chi_d$ in (17) and (18).

After N photons have been traced, and the intensities $I^{(n)}$, $I_l^{(n)}$, $I_r^{(n)}$, and $V^{(n)}$ of the N histories have been stored, we average to obtain the statistical result,

$$\bar{I} = \frac{1}{N} \sum_{n=1}^{N} I^{(n)}, \quad \bar{I}_l = \frac{1}{N} \sum_{n=1}^{N} I_l^{(n)}, \tag{19}$$

$$\bar{I}_r = \frac{1}{N} \sum_{n=1}^{N} I_r^{(n)}, \quad \bar{V} = \frac{1}{N} \sum_{n=1}^{N} V^{(n)}. \tag{20}$$

We note from (17) and (18) that $I_l + I_r = I(1 + \sin^2 \theta_d)$. This is not equal to I unless $\sin \theta_d = 0$, because if $\sin \theta_d \neq 0$ there is a component of the electric field normal to the analyzers. This component is transmitted through them independently of their orientation. Consequently, when measuring a bundle of light, in a typical experimental situation, the sum $\bar{I}_l + \bar{I}_r$ of the parallel and cross-polarized components is not equal to the total intensity $\bar{I}$, measured without polarization analyzers.

We also note that the fraction of light with the original polarization, $(\bar{I}_l - \bar{I}_r)/\bar{I}$, coincides with the common definition of the degree of polarization $\sqrt{\bar{Q}^2 + \bar{U}^2}/\bar{I}$, only for small apertures. In this case, the intensity of light that has maintained the incident polarization state, $\bar{I}_l - \bar{I}_r$, is a measure of the Stokes parameter $\bar{Q}$. In the problem that we shall consider, $\bar{U}$ is zero due to symmetry.

4.2 Splitting and Russian Roulette

One of the most widely used variance reduction methods is the combination of the so called importance splitting and Russian roulette techniques. The problem domain is divided into M subdomains with different importances $\mathcal{I}^{(m)}$ ($m = 0, \ldots, M-1$). We will assume for simplicity that $\mathcal{I}^{(m)} = c^m$ where c is a chosen positive integer. The more probable it is that a particle in a sub-domain can score at the detector, the more important is the sub-domain. Let a particle of weight W evolve between two collisions i and $i+1$, and let us denote the importance of the particle by $\mathcal{I}_i$ and $\mathcal{I}_{i+1}$ at each collision ($\mathcal{I}_i = \mathcal{I}^{(m)}$ if at collision i the photon is in sub-domain m). We form the importance ratio $\eta = \mathcal{I}_{i+1}/\mathcal{I}_i$. If $\eta = 1$ the importance of the photon does not change and its transport continues in the usual way. If $\eta > 1$ the photon has moved to a more important region and it is split into $N_{split} = \eta$ photons. The weight W of each of the split photons is divided by N_{split}. If $\eta < 1$ it has entered into a less important region, and a Russian roulette game is played where the particle has probability $p_{surv} = \eta$ of surviving. A random number ξ is drawn uniformly on the interval $(0,1)$ and compared with p_{surv}. If $\xi < p_{surv}$ the photon survives and its weight is increased by $1/p_{surv}$. Otherwise, it is eliminated. In doing this, however, a balance has to be struck. Implementing splitting generally decreases the history variance but increases the computing time per history. On the contrary, Russian roulette increases the former and decreases the latter, so that more histories can be run. The usual measure of efficiency of the Monte Carlo simulations is the figure of merit $1/\sigma^2 t$, where the variance σ^2 measures the inaccuracy of the result, and t is the mean time per history. We have seen experimentally that, for example, for a medium of optical thickness 12, the choice $c = 2$ and $M = 5$ improves the figure of merit by a factor 10. However, for optical thickness less than 8 we have not found the splitting/Russian roulette necessary. For deep penetration, experience has indicated that the best splitting results are achieved keeping the particle flux through the medium close to a flat distribution, and that there is a broad range of parameters close to the optimal choice [18]. If splitting with Russian roulette is the only variance reduction technique used, all particles in the same sub-domain m have the same weight $1/\mathcal{I}^{(m)}$, assuming that all particle histories began with weight 1.

4.3 Point Detectors

We now consider scoring at a small detector of area dA, often called a *point detector*. For each scattering event i of the n-th photon, we store the probability that the n-th scattered photon, of weight $W_i^{(n)}$, will arrive at the detector without further collisions. Let $P(\Theta, \mathcal{P})d\Omega$ be the probability of scattering into the solid angle $d\Omega$ about Θ, where Θ is the scattering angle directed to the detector, and $\mathcal{P} = (Q_0, U_0)$ denotes the liner polarization state of the photon. Taking into account the attenuation suffered by particles traveling through the medium, the probability of arriving at the detector is

$$P(\Theta, \mathcal{P})d\Omega e^{- \int_0^R \Sigma_s(l)dl} , \tag{21}$$

where R is the distance covered within the medium.

We recall that we are considering non-absorbing media. Absorption can be accounted for easily by replacing the scattering coefficient Σ_s in (21) by the total extinction coefficient $\Sigma_t = \Sigma_s + \Sigma_a$, where Σ_a is the absorption coefficient, and multiplying dI by Σ_s/Σ_t at each scattering event.

If the detector of area dA has normal vector forming the angle θ_d with respect to the line of flight, then $d\Omega = \cos\theta_d dA/R^2$, and (21) becomes

$$P(\Theta, \mathcal{P})\frac{\cos\theta_d e^{- \int_0^R \Sigma_s(l)dl}}{R^2}dA . \tag{22}$$

Since the flux is the number of particles crossing unit area per unit time, the contribution of the i-th collision to the flux at the detector is given by

$$\mathcal{F}_{Ii}^{(n)} = W_i^{(n)} P(\Theta, \mathcal{P})\frac{\cos\theta_d e^{-\Sigma_s R}}{R^2} . \tag{23}$$

Here we have evaluated the integral for a homogeneous medium. The contribution to the Stokes parameters is given by (14) times $\mathcal{F}_{Ii}^{(n)}$. In the case of a linearly polarized incident beam, the contributions to the parallel and cross-polarized components are evaluated with expressions similar to (17) and (18), respectively.

After N photons have been traced, the statistical averages are computed as

$$\bar{\mathcal{F}}_I = \frac{1}{N} \sum_{n=1}^{N} \sum_{i=1}^{S_n} \mathcal{F}_{Ii}^{(n)}, \quad \bar{\mathcal{F}}_l = \frac{1}{N} \sum_{n=1}^{N} \sum_{i=1}^{S_n} \mathcal{F}_{li}^{(n)}, \tag{24}$$

$$\bar{\mathcal{F}}_r = \frac{1}{N} \sum_{n=1}^{N} \sum_{i=1}^{S_n} \mathcal{F}_{ri}^{(n)}, \quad \bar{\mathcal{F}}_V = \frac{1}{N} \sum_{n=1}^{N} \sum_{i=1}^{S_n} \mathcal{F}_{Vi}^{(n)}, \tag{25}$$

where S_n is the number of scattering events of the n-th photon.

In spite of the simplicity of the method, the point detector estimator has a serious drawback. The expression (23) becomes infinite when a collision takes place infinitely close to the detector, giving rise to an unbounded estimator with infinite variance and consequently a low rate of convergence. However, if the detector is not in the scattering medium, a collision close to the detector is impossible and the estimator is bounded. Our point detectors are placed at a distance $d = 1$ mm from the exit surface, so we replace the R in the denominator of (23) by $(R + d/\cos\theta_d)$. Point detectors have been used widely in the literature [20–23].

Since tracing photons from each collision to the detector is time consuming, especially in the case where an object is embedded, we have also implemented a Russian roulette game for the detector. If the collision takes place in a region of importance $\mathcal{I}^{(m)}$, we introduce the probability to contribute to the flux equal to $\mathcal{I}^{(m)}/\mathcal{I}^{(M-1)} = c^{m-M+1}$. If the particle wins the game its deterministic contribution is multiplied by $\mathcal{I}^{(M-1)}/\mathcal{I}^{(m)}$ and it is stored. Otherwise, the contribution is not stored.

Finally, we note that other techniques which improve the convergence of a Monte Carlo simulation can be combined with the procedure described above. For example, Tinet et al. [22] use the statistical estimator concept to accelerate the convergence for the scalar transport equation. In their scheme, they divide the computation in two stages. The first one is the generation and storage of all the necessary information, with a random walk similar to the one explained in Sect. 4.1 but without polarization effects. However, they also compute at each scattering event, the part of the light that escapes the medium without further collisions, and they subtract it from the weight of the photon. Consequently, the photon is not allowed to leave the medium and the energy distribution within the medium is closer to the average energy distribution. In the second stage they use the information generated previously to estimate, with a point detector technique, the contribution to any quantity of interest. They report a better convergence rate than the use of the point detector alone. Its procedure is also more flexible due to the separation into two stages, which allow the post-computation of any quantity of interest related to the problem. The main drawback is the need to store the position and weight of the photons at all the scattering events, which may require a large amount of memory.

5 Numerical Results

In our numerical experiments we consider an ultra-short laser beam impinging normally at some point (x_{in}, y_{in}) on the surface $z = 0$ of a sample of dimensions $L \times L \times Z$; see Fig. 2. The laser beam is linearly polarized parallel to the (x, z) plane. The sample consists of a uniform background medium with scattering and absorption coefficients $\Sigma_{bg}^{(s)}$ and $\Sigma_{bg}^{(a)}$ respectively, containing a spherical inclusion of optical coefficients $\Sigma_{incl}^{(s)}$ and $\Sigma_{incl}^{(a)}$. See [23] for a good description of the modifications needed to handle the case where a spherical object is imbedded.

The dimensions of our sample are $10\,\mathrm{cm} \times 10\,\mathrm{cm} \times 5\,\mathrm{cm}$; see Fig. 2. The incident beam is 100% linearly polarized parallel to the (x, z) plane and enters into the medium at time $t = 0$. The incident laser positions are on the x axis 2.5 mm apart. For each laser position, two detectors are placed on the optical axis, one at the bottom, and another at the top of the sample, to measure the transmitted and reflected intensities of the light. Both detectors are at $d = 1.0\,\mathrm{cm}$ from the exit surface. The sample consists of a uniform background medium with optical parameters $\Sigma_{bg}^{(s)}$ and $\Sigma_{bg}^{(a)}$, containing a spherical inclusion of radius $R_{incl} = 5.0\,\mathrm{mm}$ and optical parameters $\Sigma_{incl}^{(s)}$ and $\Sigma_{incl}^{(a)}$. We assume that

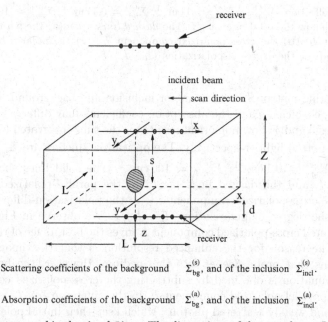

Scattering coefficients of the background $\Sigma_{bg}^{(s)}$, and of the inclusion $\Sigma_{incl}^{(s)}$.

Absorption coefficients of the background $\Sigma_{bg}^{(a)}$, and of the inclusion $\Sigma_{incl}^{(a)}$.

Fig. 2. The setup used in the simulations. The dimensions of the sample are $L \times L \times Z$. It consists of a uniform medium in which a spherical object of radius 5.0 mm, with different optical properties, is imbedded. The center of the object is at 1.6 cm from the input surface. The distance between two consecutive incident beam positions is 2.5 mm. The detectors are placed at a distance $d = 1.0\,\mathrm{cm}$ from the exit surface. The semiaperture angle of the each detector is 90°. A complete scan consists of 17 *measurements*

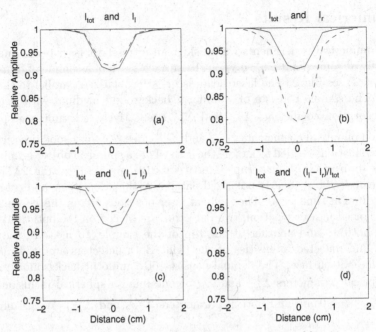

Fig. 3. Reflection. $\Sigma_{bg}^{(s)} = \Sigma_{incl}^{(s)} = 1\,\mathrm{cm}^{-1}$, $\Sigma_{bg}^{(a)} = 0.1\,\mathrm{cm}^{-1}$, $\Sigma_{incl}^{(a)} = 100\,\mathrm{cm}^{-1}$. The *solid lines* show the total intensity I. The *dashed lines* show (**a**) the parallel-polarized component I_l, (**b**) the cross-polarized component I_r, (**c**) the polarization-difference $I_l - I_r$, and (**d**) the degree of polarization $(I_l - I_r)/I$

the scattering matrix S is the same for inclusion and background. However the scattering coefficients and the absorption coefficients may differ.

Figures 3 and 4 show numerical results for the time-integrated backscattered and transmitted light, respectively. The optical parameters are $\Sigma_{bg}^{(s)} = \Sigma_{incl}^{(s)} = 1\,\mathrm{cm}^{-1}$, $\Sigma_{bg}^{(a)} = 0.1\,\mathrm{cm}^{-1}$, $\Sigma_{incl}^{(a)} = 100\,\mathrm{cm}^{-1}$. The solid lines show the total intensity I, and the dashed lines show: (a) the parallel-polarized component I_l, (b) the cross-polarized component I_r, (c) the polarization-difference $I_l - I_r$, and (d) the degree of polarization $(I_l - I_r)/I$. We observe in Fig. 3 that the backscattered cross-polarized component I_r gives the best image of the absorbing object. The reason for the enhanced resolution is that this image is formed by photons that probe deeper into the medium. However, the best image in transillumination is obtained by subtracting the cross-polarized component I_r from the parallel polarized component I_l. In the difference $I_l - I_r$, only ballistic photons and weakly scattered photons, which keep their initial polarization, are retained [4,7].

As a second example (Figs. 5 and 6) we consider a scattering object with $\Sigma_{incl}^{(s)} = 10\Sigma_{bg}^{(s)} = 10\,\mathrm{cm}^{-1}$ and $\Sigma_{incl}^{(a)} = \Sigma_{bg}^{(a)} = 0.1\,\mathrm{cm}^{-1}$. The curves in the figures have the same meaning as in Figs. 3 and 4, and the time-integrated backscattered and transmitted intensity is also shown. We obtain similar

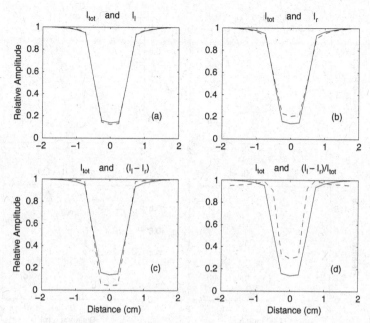

Fig. 4. Transmission. $\Sigma_{bg}^{(s)} = \Sigma_{incl}^{(s)} = 1\,\mathrm{cm}^{-1}$, $\Sigma_{bg}^{(a)} = 0.1\,\mathrm{cm}^{-1}$, $\Sigma_{incl}^{(a)} = 100\,\mathrm{cm}^{-1}$. The *solid lines* show the total intensity I. The *dashed lines* show (**a**) the parallel-polarized component I_l, (**b**) the cross-polarized component I_r, (**c**) the polarization-difference $I_l - I_r$, and (**d**) the degree of polarization $(I_l - I_r)/I$

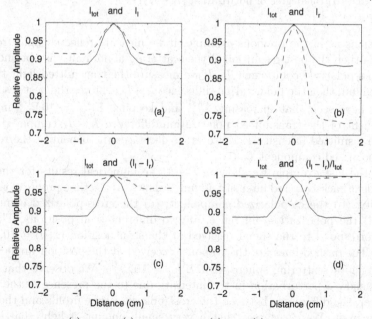

Fig. 5. Reflection. $\Sigma_{incl}^{(s)} = 10\Sigma_{bg}^{(s)} = 10\,\mathrm{cm}^{-1}$, $\Sigma_{incl}^{(a)} = \Sigma_{bg}^{(a)} = 0.1\,\mathrm{cm}^{-1}$. The *solid lines* show the total intensity I. The *dashed lines* show (**a**) the parallel-polarized component I_l, (**b**) the cross-polarized component I_r, (**c**) the polarization-difference $I_l - I_r$, and (**d**) the degree of polarization $(I_l - I_r)/I$

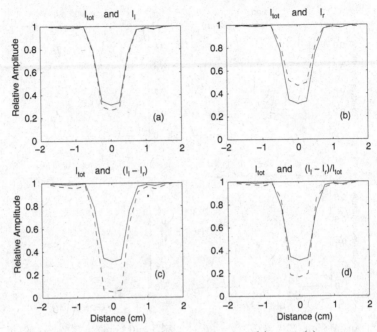

Fig. 6. Transmission. $\Sigma_{incl}^{(s)} = 10\Sigma_{bg}^{(s)} = 10\,\text{cm}^{-1}$, $\Sigma_{incl}^{(a)} = \Sigma_{bg}^{(a)} = 0.1\,\text{cm}^{-1}$. The *solid lines* show the total intensity I. The *dashed lines* show (**a**) the parallel-polarized component I_l, (**b**) the cross-polarized component I_r, (**c**) the polarization-difference $I_l - I_r$, and (**d**) the degree of polarization $(I_l - I_r)/I$

conclusions as in the previous example. If we measure reflected light from the medium then the best resolution of the scattering abnormality is obtained with the cross-polarized component I_r. If we measure the transmitted light through the medium, then the polarization-difference $I_l - I_r$ achieves the best resolution.

We have seen that increasing $\Sigma_{bg}^{(s)}$ and keeping $\Sigma_{incl}^{(s)} = 10\Sigma_{bg}^{(s)}$ increases the benefit of the transmitted polarization-difference $I_l - I_r$ compared to the total transmitted intensity I. However, it decreases the benefit of the reflected cross-polarized component I_r.

As the last example we show in Fig. 7 the numerical results for the time-dependent backscattered intensity. Figure 7 shows (a) the time profile of the total intensity, (b) the co-polarized component, (c) the cross-polarized component, and (d) the polarization-difference, at two different scan positions. The solid lines correspond to the signal received at the scan position $(x, y) = (0, 0)$; see Fig. 2. The dashed-lines are the response received at the position furthest from the target (a scattering sphere with $\Sigma_{incl}^{(s)} = 10\Sigma_{bg}^{(s)}$). We observe that all the figures show a perturbation in the intensity due to the presence of the object. However, the differences between the cross-polarized time profile and the others are apparent. We clearly see that a very small amount of light that scatters just beneath the surface $z = 0$ is present in the signal. The presence of an object that scatters light more makes the light depolarize faster. Therefore more

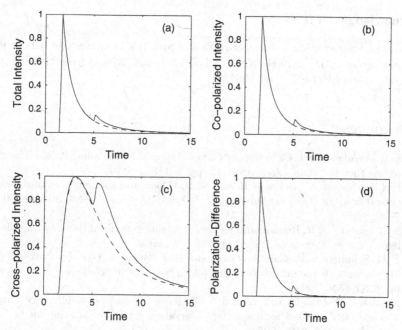

Fig. 7. Time trace of the backscattered light response. $\Sigma_{incl}^{(s)} = 10\Sigma_{bg}^{(s)} = 15\,\mathrm{cm}^{-1}$, $\Sigma_{incl}^{(a)} = \Sigma_{bg}^{(a)} = 0.1\,\mathrm{cm}^{-1}$, $L = 16\,\mathrm{cm}$, $Z = 5\,\mathrm{cm}$, $s = 1.6\,\mathrm{cm}$. (**a**) Normalized total intensity, (**b**) normalized co-polarized component, (**c**) normalized cross-polarized component, and (**d**) normalized polarization-difference. The maxima in (**b**), (**c**), and (**d**) are 1.01, 20.36 and 1.02 times smaller than the maximum in (**a**). *Solid curves* correspond to the signal received at the scan position $(x, y) = (0, 0)$, just above the center of the sphere. *Dashed curves* are the signal received at the scan position, $(x, y) = (2.8, 0)\,\mathrm{cm}$, furthest from the sphere

photons that preserve the initial polarization are present in the cross-polarized image. The fact that the cross-polarized intensity is composed of photons that have propagated deeper into the medium may be very useful for reconstruction of images when one is not interested in details at and near the surface.

6 Conclusions

We have shown that simple polarization imaging techniques can be used to enhance the resolution of objects hidden in scattering media. Polarization imaging is based on the fact that the polarization of the initial beam is preserved over large optical distances. For backscattered light, we find that the cross-polarized component shows the best resolution, while for transmitted light, images obtained by subtraction of the cross-polarized component from the co-polarized component are better. Differences between the time-dependent backscattered signal have also been presented.

Acknowledgements

The author acknowledges support from the Spanish Ministerio de Educación y Ciencia (grants no. FIS2004-03767 and FIS2007-62673) and from the European Union (grant LSHG-CT-2003-503259).

References

1. S. R. Arridge and J. C. Hebden, "Optical Imaging in Medicine: II. Modelling and reconstruction," *Phys. Med. Biol.* **42**, pp. 841–853, 1997.
2. P. J. Michielsen K., DeRaedt H. and G. N, "Computer simulation of time-resolved optical imaging of objects hidden in turbid media," *Phys. Reports* **304**, pp. 90–144, 1998.
3. S. L. Jacques, J. R. Roman, and K. Lee, "Imaging superficial tissues with polarized light," *Laser. Surg. Med.* **26**(2), pp. 119–129, 2000.
4. J. M. Schmitt, A. H. Gandjbakhche, and R. F. Bonner, "Use of polarized light to discriminate short-path photons in a multiply scattering medium," *Appl. Opt.* **31**, pp. 6535–6546, 1992.
5. M. Rowe, E. Pungh, J. Tyo, and N. Engheta, "Polarization-difference imaging: a biologically inspired technique for observation through scattering media," *Opt. Lett.* **20**, pp. 608–610, 1995.
6. S. Demos and R. Alfano, "Temporal gating in highly scattering media by the degree of optical polarization," *Opt. Lett.* **21**, pp. 161–163, 1996.
7. M. Moscoso, J. Keller, and G. Papanicolaou, "Depolarization and blurring of optical images by biological tissues," *J. Opt. Soc. Am. A* **18, No. 4**, pp. 948–960, 2001.
8. S. Chandrasekhar, *Radiative Transfer*, Oxford University Press, Cambridge, 1960.
9. A. Kim and M. Moscoso, "Chebyshev spectral methods for radiative transfer," *SIAM J. Sci. Comput.* **23, No. 6**, pp. 2075–2095, 2002.
10. B. C. Wilson and S. Jacques, "Optical reflectance and transmittance of tissues: Principles and applications," *IEEE Quant. Electron.* **26**, pp. 2186–2199, 1990.
11. J. M. Schmitt and G. Kumar, "Turbulent nature of refractive-index variations in biological tissue," *Opt. Lett.* **21**, pp. 1310–1312, 1996.
12. L. Ryzhik, G. Papanicolaou, and J. B. Keller, "Transport equations for elastic and other waves in random media," *Wave Motion* **24**, pp. 327–370, 1996.
13. A. Kim and M. Moscoso, "Influence of the relative refractive index on depolarization of multiply scattered waves," *Physical Review E* **64**, p. 026612, 2001.
14. A. Kim and M. Moscoso, "Backscattering of circularly polarized pulses," *Opt. Letters.* **27**, pp. 1589–1591, 2002.
15. B. Beauvoit, T. Kitai, and B. Chance, "Contribution to the mitochondrial compartment to the optical properties of the rat liver: a theoretical and practical approach," *Biophys. J.* **67**, pp. 2501–2510, 1994.
16. J. Mourant, J. Freyer, A. Hielscher, A. Eick, D. Shen, and T. Johnson, "Mechanisms of light scattering from biological cells relevant to noninvasive optical-tissue diagnostics," *Appl. Opt.* **37**, pp. 3586–3593, 1998.
17. M. H. Kalos and P. A. Whitlock, *Monte Carlo Methods*, New York : J. Wiley Sons, 1986.
18. E. E. Lewis and W. F. Miller Jr., *Computational Methods of Neutron Transport*, John Wiley and sons, New York, 1984.

19. G. Bal, G. Papanicolaou, and L. Ryzhik, "Probabilistic Theory of Transport Processes with Polarization," *SIAM Appl. Math.* **60**, pp. 1639–1666, 2000.
20. M. Kalos, "On the estimation of flux at a point by Monte Carlo," *Nucl. Sci. Eng.* **16**, pp. 111–117, 1963.
21. P. Bruscaglioni, G. Zaccanti, and Q. Wei, "Transmission of a pulse polarized light beam through thick turbid media: numerical results," *Appl. Opt.* **32**, pp. 6142–6150, 1993.
22. E. Tinet, S. Avrillier, and M. Tualle, "Fast semianalytical monte carlo simulation for time-resolved light propagation in turbid media," *J. Opt. Soc. Am. A* **13**, pp. 1903–1915, 1996.
23. S. Chatigny, M. Morin, D. Asselin, Y. Painchaud, and P. Beaudry, "Hybrid monte carlo for photon transport through optically thick scattering media.," *Appl. Opt.* **38**, pp. 6075–6086, 1999.

Topological Derivatives for Shape Reconstruction

Ana Carpio[1] and Maria Luisa Rapún[2]

[1] Departamento de Matemática Aplicada, Facultad de Matemáticas, Universidad Complutense de Madrid, Plaza de Ciencias 3, 28040 Madrid, Spain
ana_carpio@mat.ucm.es
[2] Departamento de Fundamentos Matemáticos de la Tecnología Aeronáutica, Escuela Técnica Superior de Ingenieros Aeronáuticos, Universidad Politécnica de Madrid, Plaza del Cardenal Cisneros 3, 28040 Madrid, Spain
marialuisa.rapun@upm.es

Summary. Topological derivative methods are used to solve constrained optimization reformulations of inverse scattering problems. The constraints take the form of Helmholtz or elasticity problems with different boundary conditions at the interface between the surrounding medium and the scatterers. Formulae for the topological derivatives are found by first computing shape derivatives and then performing suitable asymptotic expansions in domains with vanishing holes. We discuss integral methods for the numerical approximation of the scatterers using topological derivatives and implement a fast iterative procedure to improve the description of their number, size, location and shape.

1 Introduction

A huge number of applications lead to inverse scattering problems whose goal is to detect objects buried in a medium [2, 11, 17]. Such is the case of ultrasound in medicine, X-ray diffraction to retrieve information about the DNA, reflection of seismic waves in oil prospecting or crack detection in structural mechanics. A standard procedure to find information about the obstacles consists in emitting a certain type of radiation which interacts with the objects and the surrounding medium and is then measured at the detector locations. The total field, consisting of emitted, scattered and transmitted waves, solves a partial differential equation (Maxwell's equations, radiative transfer equation, equations of elasticity, Helmholtz equation...) with boundary conditions at the interface between the scatterers and the medium. The original inverse problem is the following: knowing the emitted waves and the measured patterns, find the obstacles for which the solution of the corresponding boundary value problem agrees with the measured values at the detector locations. There is a large literature on how to solve it, see [8, 11] for classical results and [2, 17] for more recent advances.

Some of the methods proposed to solve the nonlinear inverse problem are based on linear approximations such as the Born approximation [14], the Kirchhoff or physical optics approximation [36] or backpropagation principles [41]. The linear sampling method amounts to solving an equation for the contour of the unknown domain or its representative coefficients [9, 12].

Here, we will follow the variational approach. The original formulation is too restrictive [8, 30] and weaker formulations have been proposed in the variational setting [18, 29], leading to optimization problems with constraints given by boundary value problems for partial differential equations. Most optimization strategies rely on gradient principles, see for instance [33]. They deform an initial guess for the contours of the obstacles through an iterative process in such a way that the derivative of the functional to be minimized decreases. The techniques differ in the type of deformation, the representation of the contours or the way to initialize the process. The first attempts relied on classical shape optimization using a small perturbation of the identity [27, 31, 38, 44, 52]. However, this strategy does not allow for topological changes in the obstacle, i.e., the number of components has to be known from the beginning. This problem was solved by introducing deformations inspired in level-set methods [37, 50]. In this way, contours may be created or destroyed during the iterative process. Nevertheless, level-set based iterative methods may be rather slow unless a good initial guess of the scatterers is available.

Topological derivative methods arise as an efficient alternative to solve inverse scattering problems [53]. There is a growing literature on the subject, see for instance [2, 18, 26] for scattering problems in acoustics, [21, 25] for the elasticity case and [17, 39, 49] for problems involving electromagnetic waves. In scattering problems, the topological derivative of the cost functional under study provides a good first guess of the number and location of obstacles, as shown in [18, 25]. This guess may be improved using iterative schemes entirely based on topological derivatives, as discussed in [4, 21]. Hybrid level set-topological derivative methods have been suggested in [3].

The paper is organized as follows. Section 2 discusses topological derivative methods when the interaction between the scatterers, the medium and the incident radiation is described by Helmholtz problems. We describe the forward scattering model and reformulate the inverse problem as a constrained optimization problem. Then, we recall the expressions for the topological derivatives of the resulting functionals with different boundary conditions. Reconstructions of objects obtained using topological derivatives with synthetic data are shown in Sect. 3. Good approximations for the number, location and size of the scatterers are found. Section 4 presents an iterative scheme which improves the description of the shape of the obstacles in a few iterations. Section 5 gives some details on integral methods for the numerical approximation of topological derivatives. In Sect. 6, the analytic procedure to compute explicit expressions for the topological derivatives in Helmholtz problems is explained. Extensions to detection of objects by means of acoustic waves in elastodynamic problems are presented in Sect. 7. Finally, conclusions and directions for future work are discussed in Sect. 8.

2 Helmholtz Equations

2.1 The Forward and Inverse Scattering Problems

Let us assume we have a certain medium where a finite number of obstacles $\Omega_{i,j}$, $j = 1, \ldots, d$ are buried (the subscript i stands for interior), as in Fig. 1. A classical procedure to locate the unknown obstacles, consists in illuminating the medium with a particular type of wave (electromagnetic, acoustic, pressure...). The incident radiation interacts with the medium and the obstacles, is reflected and then measured at a set of receptors Γ_{meas}, placed far enough from the objects. Knowing the emitted and measured wavefields, it might be possible to find the obstacles. Many detecting devices follow this principle: radars, sonars, lidars, scanners. ...

The total field $u = u_{inc} + u_{sc}$ formed by the incident and the scattered wave satisfies a boundary problem for some partial differential equation (Maxwell's equations, elasticity equations ...). This section develops techniques to detect objects when the governing partial differential equation is a Helmholtz equation. This happens in simple geometries for either electromagnetic (TM, TE polarized waves) or acoustic waves [17, 18, 24].

Let us make precise the setting we will be working in. The recepting set Γ_{meas} may be a curve in simple tests, or a number of discrete points at which receptors are located in more realistic situations. To simplify, we take the medium to be $\mathbb{R}^2$ and the incident radiation to be a plane wave $u_{inc}(\mathbf{x}) = \exp(\imath \lambda_e \mathbf{x} \cdot \mathbf{d})$ with wave number λ_e and propagation direction $\mathbf{d}$, $|\mathbf{d}| = 1$. The obstacle $\Omega_i \subset \mathbb{R}^2$ is an open bounded set with smooth boundary $\Gamma := \partial \Omega_i$ but has no assumed connectivity. There may be an unknown number of isolated components: $\Omega_i = \cup_{j=1}^d \Omega_{i,j}$ in which $\Omega_{i,j}$ are open connected bounded sets satisfying $\overline{\Omega}_{i,l} \cap \overline{\Omega}_{i,j} = \emptyset$ for $l \neq j$.

When the incident wave penetrates the scatterer ('penetrable' obstacle), part of the wave is reflected and the rest is transmitted inside. Transmission conditions are imposed at the interface between the obstacle and the surrounding medium.

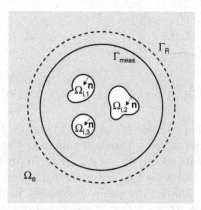

Fig. 1. Geometry of the problem: $\Omega_i = \sum_{j=1}^d \Omega_{i,j}$ and $\Omega_e = \mathbb{R}^2 \setminus \overline{\Omega}_i$

The total wavefield $u = u_{inc} + u_{sc}$ in $\Omega_e = \mathbb{R}^2 \setminus \overline{\Omega}_i$ (the subscript e stands for exterior) and the transmitted wave field $u = u_{tr}$ in Ω_i solve

$$\begin{cases} \Delta u + \lambda_e^2 u = 0, & \text{in } \Omega_e, \\[2mm] \alpha_i \Delta u + \lambda_i^2 u = 0, & \text{in } \Omega_i, \\[2mm] u^- - u^+ = 0, & \text{on } \Gamma, \\[2mm] \alpha_i \partial_\mathbf{n} u^- - \partial_\mathbf{n} u^+ = 0, & \text{on } \Gamma, \\[2mm] \lim_{r \to \infty} r^{1/2} \left(\partial_r(u - u_{inc}) - \imath\lambda_e(u - u_{inc}) \right) = 0, & r = |\mathbf{x}|. \end{cases} \qquad (1)$$

We select the unit normal vector $\mathbf{n}$ pointing outside Ω_e. $\partial_\mathbf{n}$ stands for the normal derivative at the interface and ∂_r is the radial derivative. We denote by u^+ and u^- the limits of u from the exterior and interior of Ω_i respectively. To simplify, the parameters $\lambda_e, \lambda_i, \alpha_i$ are taken to be constant, real and positive. With this assumption, we can impose the standard Sommerfeld radiation condition on the propagation of the scattered field $u_{sc} = u - u_{inc}$ at infinity, which implies that only outgoing waves are allowed.

General conditions on the parameters λ_e, λ_i, α_i ensuring existence and uniqueness of a solution $u \in H^1_{loc}(\mathbb{R}^2)$ for this problem can be found for instance in [13, 32, 34, 54, 55].

Alternatively, we might reformulate (1) as a boundary value problem for the scattered and the transmitted fields with transmission conditions $u_{tr} - u_{sc} = u_{inc}$ and $\alpha_i \partial_\mathbf{n} u_{tr} - \partial_\mathbf{n} u_{sc} = \partial_\mathbf{n} u_{inc}$. This is usually done for numerical purposes. However, (1) simplifies the later computation of shape and topological derivatives for this problem. The parameter α_i is not scaled out in the second equation for the same reason.

When the scatterer is opaque to the incident radiation ('sound hard' obstacle), there is no transmitted wave and a Neumann boundary condition is imposed at the interface:

$$\begin{cases} \Delta u + \lambda_e^2 u = 0, & \text{in } \Omega_e, \\[2mm] \partial_\mathbf{n} u = 0, & \text{on } \Gamma, \\[2mm] \lim_{r \to \infty} r^{1/2} \left(\partial_r(u - u_{inc}) - \imath\lambda_e(u - u_{inc}) \right) = 0, & r = |\mathbf{x}|. \end{cases} \qquad (2)$$

The forward problem consists in solving (1) (resp. (2)) when the obstacles Ω_i and the incident field u_{inc} are known. Evaluating the solution at Γ_{meas}, we recover the measured pattern: $u|_{\Gamma_{meas}} = u_{meas}(\mathbf{d})$. In practical experiments, the total field u is known on the set of receptors Γ_{meas} for different directions of the incident wave $\mathbf{d}_j$. The inverse problem consists in finding the shape and structure of the obstacle Ω_i such that the solution of the forward problem (1) (resp. (2)) equals the measured values $u_{meas}(\mathbf{d}_j)$ at the receptors.

This inverse problem is nonlinear with respect to Ω_i and strongly ill-posed [8]. Given arbitrary data u_{meas}, an associated scatterer Ω_i may not exist. When it exists, it may not depend continuously on u_{meas}. It is well-known that the obstacles are uniquely determined by the far-field pattern of the scattered wave for all incident directions and one fixed wave number λ_e (see the relevant paper [30] and the latter work [23] where the proofs in [30] for the transmission problem were simplified). Therefore, by an analyticity argument (see [11]), if Γ_{meas} is a circumference, then the values of the total wave on Γ_{meas} for all incident waves determine uniquely the obstacles. The question of uniqueness without any a priori knowledge about the location of the obstacles for just one (or a finite number of) incident plane waves is still an open problem.

Different strategies to regularize the inverse problem have been proposed in the literature. We resort here to the variational approach: look for a domain Ω_i which minimizes an error in some sense. This leads to a constrained optimization problem: minimize

$$J(\Omega_i) := \frac{1}{2} \int_{\Gamma_{meas}} |u - u_{meas}|^2 dl, \tag{3}$$

where u is the solution of the forward problem (1) (resp. (2)) and u_{meas} the measured total field on Γ_{meas}. This functional depends on the design variable Ω_i through the boundary value problem, which acts as a constraint. When several measurements corresponding to different directions of illumination $\mathbf{d}_j$ are available, the optimization problem becomes: minimize

$$J(\Omega_i) := \frac{1}{2} \sum_{j=1}^{N} \int_{\Gamma_{meas}} |u^j - u_{meas}^j|^2 dl, \tag{4}$$

in which u^j are the solutions of N forward problems with incident waves $u_{inc}^j(\mathbf{x}) = \exp(\imath \lambda_e \mathbf{x} \cdot \mathbf{d}_j)$. We have now N constraints.

2.2 Topological Derivatives Applied to Inverse Scattering

Different notions of derivative have been introduced for shape functionals. The topological derivative of a functional measures its sensitivity when infinitesimal holes are removed from a domain. When applied to functionals associated to inverse scattering problems such as (3), the topological derivative provides information on the location of the scatterers: Regions where the topological derivative takes on large negative values will be identified with places where an object should be located.

The standard formal definition is the following. Let us consider a small ball $B_\varepsilon(\mathbf{x}) = B(\mathbf{x}, \varepsilon)$, $\mathbf{x} \in \mathcal{R} \subset \mathbb{R}^2$, and the domain $\mathcal{R}_\varepsilon := \mathcal{R} \setminus \overline{B}_\varepsilon(\mathbf{x})$. The topological derivative of $\mathcal{J}(\mathcal{R})$ is a scalar function of $\mathbf{x} \in \mathcal{R}$ defined as

$$D_T(\mathbf{x}, \mathcal{R}) := \lim_{\varepsilon \to 0} \frac{\mathcal{J}(\mathcal{R}_\varepsilon) - \mathcal{J}(\mathcal{R})}{\mathcal{V}(\varepsilon)}. \tag{5}$$

The scalar function $\mathcal{V}(\varepsilon)$ is chosen in such a way that the limit (5) exists, is finite and does not vanish. $\mathcal{V}(\varepsilon)$ is usually related to the measure of the ball. In our case, $\mathcal{V}(\varepsilon) = -\pi\varepsilon^2$. In general, it will be a decreasing and negative function, satisfying $\lim_{\varepsilon \to 0} \mathcal{V}(\varepsilon) = 0$. The value of the limit (5) depends on the partial differential equation we use as a constraint and on the boundary conditions we impose on the boundary of the hole. This limit also depends on the kind of hole we are removing. It would change slightly for shapes different from a ball.

The relationship

$$J(\mathcal{R}_\varepsilon) = J(\mathcal{R}) + \mathcal{V}(\varepsilon)\, D_T(\mathbf{x}, \mathcal{R}) + o(\mathcal{V}(\varepsilon)),$$

allows to establish a link with the shape derivative (see [18]), which is useful to obtain explicit expressions. The shape derivative of a functional $J(\mathcal{R})$ along a vector field $\mathbf{V}(\mathbf{z})$ is defined as

$$DJ(\mathcal{R}) \cdot \mathbf{V} := \frac{d}{d\tau} J(\varphi_\tau(\mathcal{R}))\Big|_{\tau=0},$$

where φ_τ are deformations along the field $\mathbf{V}$:

$$\varphi_\tau(\mathbf{z}) := \mathbf{z} + \tau\mathbf{V}(\mathbf{z}) =: \mathbf{z}_\tau, \qquad \mathbf{z} \in \mathbb{R}^2.$$

To compute topological derivatives, we select a smooth vector field of the form

$$\mathbf{V}(\mathbf{z}) = V_n\, \mathbf{n}(\mathbf{z}), \qquad \mathbf{z} \in \partial B_\varepsilon(\mathbf{x}),$$

with constant $V_n < 0$ and extend $\mathbf{V}$ to $\mathbb{R}^2$ in such a way that it vanishes out of a neighborhood of ∂B_ε. Then,

$$D_T(\mathbf{x}, \mathcal{R}) = \lim_{\varepsilon \to 0} \left(\frac{1}{\mathcal{V}'(\varepsilon)|V_n|} \frac{d}{d\tau} J(\varphi_\tau(\mathcal{R}_\varepsilon))\Big|_{\tau=0} \right), \qquad \mathbf{x} \in \mathcal{R}, \qquad (6)$$

where $\mathcal{V}'(\varepsilon)$ is the derivative of the function $\mathcal{V}(\varepsilon)$.

The value of the topological derivative depends on the set $\mathcal{R}$. When there is no a priori information on the scatterers, one computes the topological derivative of the functional (3) setting $\mathcal{R} = \mathbb{R}^2$ and $\Omega_i = \emptyset$ to obtain a first guess of the number and location of the obstacles. As we will show later, a first approximation to the obstacles is found by selecting a domain Ω_i^1 where $D_T(\mathbf{x}, \mathbb{R}^2)$ is smaller than a certain large negative value. Setting now $\mathcal{R} = \mathbb{R}^2 \setminus \overline{\Omega}_i^1$, we may compute $D_T(\mathbf{x}, \mathbb{R}^2 \setminus \overline{\Omega}_i^1)$ and select a new approximation Ω_i^2. Iterating this procedure, we get a sequence of approximations Ω_i^k. At each step, we remove from $\mathbb{R}^2$ a set Ω_i^k in which the topological derivative is large and negative, the value of $D_T(\mathbf{x}, \mathbb{R}^2 \setminus \overline{\Omega}_i^k)$ becomes smaller and Ω_i^k gets closer to the true scatterer Ω_i.

Using (6), we find explicit expressions for the required topological derivatives of (3). We collect the main results about Helmholtz transmission and Neumann problems in two theorems whose proofs are rather technical and will be postponed till Sect. 6. Let us first consider the topological derivative in the whole space $\mathbb{R}^2$.

Theorem 2.1. *Depending on the boundary conditions at the interface between the medium and the obstacles, the topological derivative of the cost functional (3) in $\mathcal{R} = \mathbb{R}^2$ takes the following form:*

- *General transmission problem (penetrable obstacle)*

$$D_T(\mathbf{x}, \mathbb{R}^2) = \mathrm{Re}\left[\frac{2(1-\alpha_i)}{1+\alpha_i}\nabla u(\mathbf{x})\nabla \overline{p}(\mathbf{x}) + (\lambda_i^2 - \lambda_e^2)\, u(\mathbf{x})\overline{p}(\mathbf{x})\right], \qquad (7)$$

- *Transmission problem with $\alpha_i = 1$*

$$D_T(\mathbf{x}, \mathbb{R}^2) = \mathrm{Re}\left[(\lambda_i^2 - \lambda_e^2)\, u(\mathbf{x})\, \overline{p}(\mathbf{x})\right], \qquad (8)$$

- *Neumann problem (rigid obstacle)*

$$D_T(\mathbf{x}, \mathbb{R}^2) = \mathrm{Re}\left[2\nabla u(\mathbf{x})\nabla \overline{p}(\mathbf{x}) - \lambda_e^2 u(\mathbf{x})\overline{p}(\mathbf{x})\right], \qquad (9)$$

for $\mathbf{x} \in \mathbb{R}^2$. In all cases, u and p solve forward and adjoint problems with $\Omega_i = \emptyset$. The solution u of the forward problem is the incident wave $u_{inc}(\mathbf{x})$. The adjoint state p solves

$$\begin{cases} \Delta p + \lambda_e^2 p = (u_{meas} - u)\delta_{\Gamma_{meas}}, & \text{in } \mathbb{R}^2, \\[2mm] \lim_{r\to\infty} r^{1/2}\left(\partial_r p + \imath\lambda_e\, p\right) = 0, \end{cases} \qquad (10)$$

where δ_{meas} is the Dirac delta function on the sampling interface Γ_{meas}.

Theorem 2.1 holds regardless of the structure of the incident wave. It may be a plane wave $u_{inc}(\mathbf{x}) = \exp(\imath\lambda_e\, \mathbf{d} \cdot \mathbf{x})$, or a different type of source, as in Sect. 3. Notice that the forward and adjoint solutions needed for the computation of topological derivatives are independent of the boundary conditions at the interface. However, they affect the final expression when calculating the limit $\varepsilon \to 0$.

Let us see how (7)–(9) change when we compute the topological derivative of (3) in a domain with a hole $\mathcal{R} = \mathbb{R}^2 \setminus \overline{\Omega}$, Ω being an open bounded set, not necessarily connected. The new expression is formally identical, but the adjoint and forward problems are solved in $\Omega_e = \mathbb{R}^2 \setminus \overline{\Omega}$ and $\Omega_i = \Omega$, with transmission or Neumann boundary conditions at the interface $\Gamma = \partial\Omega$. Now, the boundary conditions not only determine the limit value but also affect the forward and adjoint fields u and p.

Theorem 2.2. *Depending on the boundary conditions at the interface, the topological derivative of the cost functional (3) in $\mathcal{R} = \mathbb{R}^2 \setminus \overline{\Omega}$ is given by*

- *General transmission problem*

$$D_T(\mathbf{x}, \mathbb{R}^2 \setminus \overline{\Omega}) = \mathrm{Re}\left[\frac{2(1-\alpha_i)}{1+\alpha_i}\nabla u(\mathbf{x})\nabla \overline{p}(\mathbf{x}) + (\lambda_i^2 - \lambda_e^2)\, u(\mathbf{x})\overline{p}(\mathbf{x})\right], \qquad (11)$$

where u solves the forward transmission problem (1) *with $\Omega_i = \Omega$ and p solves the adjoint transmission problem*

$$\begin{cases} \Delta p + \lambda_e^2 p = (u_{meas} - u)\,\delta_{\Gamma_{meas}}, & \text{in } \Omega_e, \\[2mm] \alpha_i \Delta p + \lambda_i^2 p = 0, & \text{in } \Omega_i, \\[2mm] p^- - p^+ = 0, & \text{on } \Gamma, \\[2mm] \alpha_i \partial_{\mathbf{n}} p^- - \partial_{\mathbf{n}} p^+ = 0, & \text{on } \Gamma, \\[2mm] \lim_{r \to \infty} r^{1/2}\left(\partial_r p + \imath\lambda_e p\right) = 0, & \end{cases} \qquad (12)$$

- *Transmission problem with $\alpha_i = 1$*

$$D_T(\mathbf{x}, \mathbb{R}^2 \setminus \overline{\Omega}) = \text{Re}\left[(\lambda_i^2 - \lambda_e^2)\,u(\mathbf{x})\,\overline{p}(\mathbf{x})\right], \qquad (13)$$

where u and p solve the forward and adjoint transmission problems (1) *and* (12) *with $\Omega_i = \Omega$ and $\alpha_i = 1$,*
- *Neumann problem*

$$D_T(\mathbf{x}, \mathbb{R}^2 \setminus \overline{\Omega}) = \text{Re}\left[2\nabla u(\mathbf{x})\nabla\overline{p}(\mathbf{x}) - \lambda_e^2 u(\mathbf{x})\overline{p}(\mathbf{x})\right], \qquad (14)$$

where u solves the Neumann problem (2) *with $\Omega_i = \Omega$ and p solves the adjoint Neumann problem*

$$\begin{cases} \Delta p + \lambda_e^2 p = (u_{meas} - u)\,\delta_{\Gamma_{meas}}, & \text{in } \Omega_e, \\[2mm] \partial_{\mathbf{n}} p = 0, & \text{on } \Gamma, \\[2mm] \lim_{r \to \infty} r^{1/2}\left(\partial_r p + \imath\lambda_e p\right) = 0, & \end{cases} \qquad (15)$$

for $\mathbf{x} \in \mathbb{R}^2 \setminus \overline{\Omega}$.

The Neumann case is recovered taking the limit $\alpha_i \to 0$, as expected [47].

Analogous formulae hold in three dimensions. The gradient term takes the form $\nabla u \mathcal{P} \nabla \overline{p}$, where $\mathcal{P}$ depends on the solutions of auxiliary transmission problems in a fixed ball, which change with dimension, see [24]. Inhomogeneous problems with variable parameters can be handled in a similar way, see [4].

The next sections present several tests illustrating the ability of topological derivative based methods to locate scatterers.

3 Numerical Computation of Topological Derivatives

In this section, we reconstruct scatterers buried in a medium by schemes based on the computation of topological derivatives. Several numerical experiments illustrate the main advantages and limitations of the method. We explain the technique in the model case of the Helmholtz transmission problem discussed in Sect. 2. The method can be extended to elasticity problems using the results in Sect. 7.

The idea is to plot $D_T(\mathbf{x}, \mathbb{R}^2)$ on a grid of points to obtain a map showing the regions where $D_T(\mathbf{x}, \mathbb{R}^2)$ is large or small. Domains where the topological derivative takes large negative values indicate the possible location of an object. Notice that the true scatterers enter the adjoint problem through the measured data u_{meas}. Our tests here use artificial data generated by solving the forward transmission problem for Ω_i and evaluating its solution at Γ_{meas}. Alternatively, real scattering data might be used.

In practice, one generates one or several different incident fields and measures the resulting total field not on a whole interface Γ_{meas} but on a finite set of receptors $\{\mathbf{x}_1, \ldots, \mathbf{x}_M\}$. We will only consider two kinds of incident fields: plane waves, which are modeled by the functions

$$u_{inc}(\mathbf{x}) := \exp(\imath \lambda_e \mathbf{x} \cdot \mathbf{d}), \qquad |\mathbf{d}| = 1, \tag{16}$$

where $\mathbf{d}$ is the propagation direction, and point sources, which are described by the functions

$$u_{inc}(\mathbf{x}) := \begin{cases} \dfrac{\imath}{4} H_0^{(1)}(\lambda_e |\mathbf{x} - \mathbf{z}|), & \text{in 2D}, \\[2mm] \dfrac{\exp(\imath \lambda_e |\mathbf{x} - \mathbf{z}|)}{4\pi |\mathbf{x} - \mathbf{z}|}, & \text{in 3D}, \end{cases} \tag{17}$$

where $\mathbf{z}$ is the focus and $H_0^{(1)}$ is the Hankel function of the first kind and order zero.

Our goal is to minimize the functional (compare with (3))

$$J(\Omega_i) := \frac{1}{2} \sum_{k=1}^{M} |u(\mathbf{x}_k) - u_{meas}(\mathbf{x}_k)|^2, \tag{18}$$

where $u_{meas}(\mathbf{x}_k)$ is the measured value of the total field at the k-th receptor $\mathbf{x}_k$ and u is the solution to the Helmholtz transmission problem associated to the domain Ω_i for an incident wave of the form (16) or (17). It is common in practice to know the total fields for several incident waves, that is, for incident fields corresponding to different propagation directions $\mathbf{d}_1, \ldots, \mathbf{d}_N$, or for different point sources located at $\mathbf{z}_1, \ldots, \mathbf{z}_N$. In this case, we minimize the cost functional (which is the discrete version of (4))

$$J(\Omega_i) := \frac{1}{2} \sum_{j=1}^{N} \sum_{k=1}^{M} |u^j(\mathbf{x}_k) - u_{meas}^j(\mathbf{x}_k)|^2, \tag{19}$$

where u^j is the solution to the forward transmission problem for the planar incident field in the direction $\mathbf{d}_j$ or the field generated at the source point $\mathbf{z}_j$, and $u_{meas}^j(\mathbf{x}_k)$ is the measured total field at the observation point $\mathbf{x}_k$.

In this section, we focus on the practical situation in which one does not have any information about the number of obstacles or its location and therefore starts by assuming that no obstacles exist. We set $\Omega_i = \emptyset$ and $\Omega_e = \mathbb{R}^n$, with $n = 2$ or $n = 3$. Then, the solution to the direct problem is simply the incident wave, given by the explicit formulae (16) or (17). The adjoint problem for $\overline{p}$ is:

$$\begin{cases} \Delta \overline{p} + \lambda_e^2 \overline{p} = \sum_{k=1}^{M} \overline{(u_{meas}(\mathbf{x}_k) - u(\mathbf{x}_k))} \delta_{\mathbf{x}_k}, \\ \lim_{r \to \infty} r^{(n-1)/2}(\partial_r \overline{p} - \imath \lambda_e \overline{p}) = 0. \end{cases} \qquad (20)$$

An explicit formula for $\overline{p}$ is obtained using the outgoing fundamental solutions of the Helmholtz equation $\Delta u + \lambda^2 u = 0$

$$\phi_\lambda(\mathbf{x}, \mathbf{y}) := \begin{cases} \dfrac{\imath}{4} H_0^{(1)}(\lambda|\mathbf{x} - \mathbf{y}|), & \text{in 2D}, \\ \dfrac{\exp(\imath \lambda|\mathbf{x} - \mathbf{y}|)}{4\pi|\mathbf{x} - \mathbf{y}|}, & \text{in 3D}, \end{cases} \qquad (21)$$

which satisfy the Sommerfeld radiation condition at infinity and $\Delta u + \lambda^2 u = \delta_{\mathbf{y}}$. Then, the solution of the adjoint problem (20) is

$$\overline{p}(\mathbf{x}) = \sum_{k=1}^{M} \overline{(u_{meas}(\mathbf{x}_k) - u(\mathbf{x}_k))} \, \phi_{\lambda_e}(\mathbf{x}, \mathbf{x}_k). \qquad (22)$$

We are ready now to compute the topological derivative of the cost functionals. In the 2D case, the topological derivative of (18) is

$$D_T(\mathbf{x}, \mathbb{R}^2) = \text{Re}\left[\frac{2(1 - \alpha_i)}{1 + \alpha_i} \nabla u(\mathbf{x}) \nabla \overline{p}(\mathbf{x}) + (\lambda_i^2 - \lambda_e^2) u(\mathbf{x}) \overline{p}(\mathbf{x})\right],$$

formally identical to the result in Theorem 2.1, but now the adjoint field is given by (22). The topological derivative of the cost functional (19) is

$$D_T(\mathbf{x}, \mathbb{R}^2) = \sum_{j=1}^{N} \text{Re}\left[\frac{2(1 - \alpha_i)}{1 + \alpha_i} \nabla u^j(\mathbf{x}) \nabla \overline{p}^j(\mathbf{x}) + (\lambda_i^2 - \lambda_e^2) u^j(\mathbf{x}) \overline{p}^j(\mathbf{x})\right], \qquad (23)$$

with p^j given by (22) with u and u_{meas} replaced by u^j and u_{meas}^j, respectively. In 3D, the factor $2(1 - \alpha_i)/(1 + \alpha_i)$ should be replaced by $3(1 - \alpha_i)/(2 + \alpha_i)$ in the case of spherical holes (see [4, 24]).

If we had a priori information of the number, size and location of the obstacles, we might set $\Omega_i \neq \emptyset$. In this case, u is not just the incident wave but the solution to the Helmholtz transmission problem for Ω_i. The adjoint problem is also slightly modified, and again a transmission problem has to be solved, as stated in Theorem 2.2. This case can be seen as the second step in an iterative method based on topological derivatives and will be studied in Sect. 4.

Let us investigate the performance of the method when no a priori information is available (that is, starting with $\Omega_i = \emptyset$) for scattering problems of time-harmonic acoustic waves. A time-harmonic incident field has the form

$$U_{inc}(\mathbf{x}, t) = \mathrm{Re}(u_{inc}(\mathbf{x})e^{-\iota\omega t}),$$

for a frequency $\omega > 0$. The complex amplitude of the incident field, $u_{inc}(\mathbf{x})$, will be a planar wave, given by (16) or a point source, given by (17). Asymptotically, it generates a time-harmonic response $U(\mathbf{x}, t) = \mathrm{Re}(u(\mathbf{x})e^{-\iota\omega t})$. The space-dependent amplitudes of the scattered wave u in Ω_e and the transmitted wave $u = u_{tr}$ in Ω_i solve the Helmholtz transmission problem (1) with wave numbers $\lambda_k = \sqrt{\omega \rho_k}$, $k = i, e$, where ρ_k is the density of the material occupying the domain Ω_k.

We start by describing a simple 2D example in a material with a single obstacle inside. The experiment consists in measuring the total field on sampling points located on a circumference, for planar incident fields of the form (16) with directions $\mathbf{d}_j := (\cos\theta_j, \sin\theta_j)^\top$. In our numerical experiments, the propagation angles θ_j are always uniformly distributed on the whole interval $[0, 2\pi)$ or on a general interval $[\beta_1, \beta_2] \subset [0, 2\pi)$. We take $\alpha_i = 1$, and densities ρ_e, ρ_i satisfying

$$\frac{\lambda_e}{\lambda_i} = \sqrt{\frac{\rho_e}{\rho_i}} = 5.$$

The composite material is excited at different frequencies ω corresponding to wave numbers λ_e and λ_i satisfying the ratio above.

In Fig. 2 we represent the topological derivative (23) obtained for measurements of the total wavefield at the 14 sampling points marked with the symbol

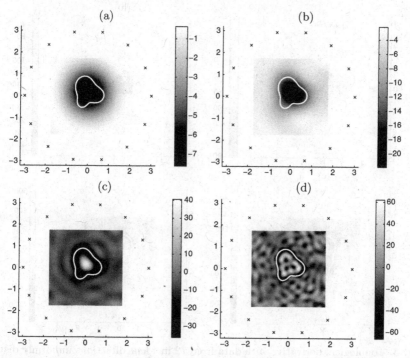

Fig. 2. Topological derivative with data from 24 incident directions uniformly distributed in $[0, 2\pi)$ (every 15°) at the 14 receptors ('×'). (a) $\lambda_e = 1.25$, $\lambda_i = 0.25$, $\alpha_i = 1$ (b) $\lambda_e = 2.5$, $\lambda_i = 0.5$, $\alpha_i = 1$ (c) $\lambda_e = 5$, $\lambda_i = 1$, $\alpha_i = 1$ (d) $\lambda_e = 10$, $\lambda_i = 2$, $\alpha_i = 1$

'×' for 24 incident planar waves with angles in $[0, 2\pi)$. Incident fields were generated at different frequencies ω, increasing its values from (a) to (d). In the four simulations we detect the presence of just one obstacle. Moreover, we guess its location and approximate size. For some of the frequencies, the approximate shape of the obstacle is also quite well recovered. Low and high frequencies provide different information about the obstacle. At low frequencies (see Fig. 2a,b), the lowest values of the topological derivative are reached inside the obstacle, located in a region of circular or elliptical shape. This allows for good reconstructions of the location and size of the obstacle, but gives little information about its shape. On the other hand, for high frequencies (see Fig. 2c,d), the lowest values of the topological derivative are concentrated in a small annular region about the boundary. Increasing the frequency, the annular region becomes thinner. With this information, one can predict more accurately the exact shape of the obstacle. For the same amount of data, the topological derivative is less smooth at high frequencies (in the sense that it shows strong oscillations) than at low frequencies. This fact makes probably more difficult to distinguish the obstacle.

Considering a larger number of receptors and/or a larger number of incident waves, the reconstructions for the low frequencies improve only a bit, but the quality for the higher ones is considerably better, as can be seen in Fig. 3.

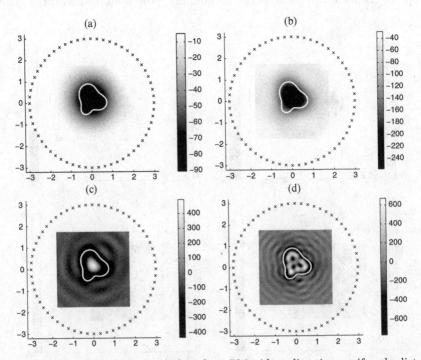

Fig. 3. Topological derivative with data from 72 incident directions uniformly distributed in $[0,2\pi)$ (every $5°$) at the 60 receptors ('×'). (a) $\lambda_e = 1.25$, $\lambda_i = 0.25$, $\alpha_i = 1$ (b) $\lambda_e = 2.5$, $\lambda_i = 0.5$, $\alpha_i = 1$ (c) $\lambda_e = 5$, $\lambda_i = 1$, $\alpha_i = 1$ (d) $\lambda_e = 10$, $\lambda_i = 2$, $\alpha_i = 1$

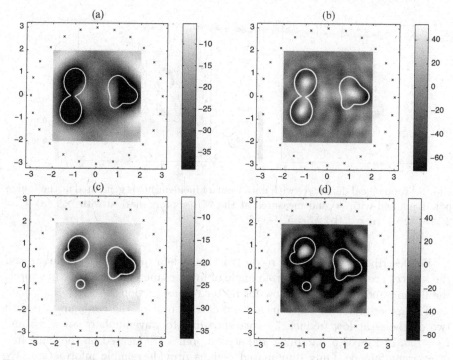

Fig. 4. Topological derivative with data from 24 incident directions uniformly distributed in $[0,2\pi)$ (every $15°$) at the 24 receptors ('×'). (a, c) $\lambda_e = 2.5$, $\lambda_i = 0.5$, $\alpha_i = 1$ (b, d) $\lambda_e = 5$, $\lambda_i = 1$, $\alpha_i = 1$

Let us analyze the performance of the method when dealing with more complex geometries. We show results for composite materials with two and three inclusions in Fig. 4a,b and Fig. 4c,d respectively. The topological derivative is computed for a low frequency in Fig. 4a,c and for a high one in Fig. 4b,d. The quality of the reconstruction of the boundaries is acceptable near the observation points. The reconstruction of other parts of the boundaries is made more difficult by the influence of each obstacle on the others. Notice that the overall reconstruction worsens in comparison with the case of a single obstacle. In Fig. 4a,b we seem to detect three inclusions instead of two. The small defect is almost undistinguishable in Fig. 4c,d. Furthermore, at low frequencies neither shapes nor locations are well recovered.

In the previous examples we considered planar incident waves. We explore now the choice of point sources. We repeat the experiment for the configuration with two obstacles, keeping the same frequencies and receptors as in Fig. 4a,b, but working now with incident fields generated at 24 uniformly distributed points on the circumference. The location of the source points is represented in Fig. 5 with a '•' symbol. The quality of the reconstruction is similar for point sources and planar waves.

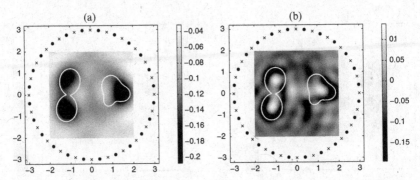

Fig. 5. Topological derivative with data from 24 incident fields generated at the source points marked with '•' and measured at the 24 receptors marked with '×'. (a) $\lambda_e = 2.5$, $\lambda_i = 0.5$, $\alpha_i = 1$ (b) $\lambda_e = 5$, $\lambda_i = 1$, $\alpha_i = 1$

We describe now numerical reconstructions when the observation points and the sources are distributed over an angle of limited aperture. As one can expect, the illuminated regions will allow for better reconstructions.

The experiments in Fig. 6 have been carried out for the configuration with two obstacles at low frequencies corresponding to wavenumbers $\lambda_e = 2.5$ and $\lambda_i = 0.5$. In Fig. 6a,b, sampling and source points are located on the same half of a circumference. Only illuminated regions provide reliable information. We recover the upper part of the obstacles in Fig. 6a and their lower part in Fig. 6b. In Fig. 6c,d source and observation points are located in complementary half circumferences, covering the whole circumference. In contrast with the results shown in Fig. 5, we predict the occurrence of only two obstacles, but their approximate sizes and shapes are poorly reproduced.

We have also investigated what happens when the source and sampling points are located on a half circumference close to one obstacle but far from the other, see Fig. 6e. As one can expect, we clearly distinguish the presence of the nearest obstacle to the sources and receptors. Its size and shape is quite well recovered although its location is displaced to the left (compare with the reconstruction in Fig. 5 where this obstacle seems to be split into two different ones, but the reconstruction of its location is more accurate). The object located in the shadow region is completely ignored by the topological derivative. Finally, in Fig. 6f, data are sampled in a quarter of the circumference and only the inferior part of the illuminated obstacle is detected.

Our numerical experiments are restricted here to Helmholtz transmission problems with $\alpha_i = 1$ because formula (8) is slightly simpler to implement. Numerical computations for general transmission problems with $\alpha_i \neq 1$ yield analogous conclusions. The interested reader can find some examples in [4]. For tests with the Neumann problem we refer to [18].

We end this section by inspecting the 3D case. Our goal is to recover the shape, size and location of a sphere when the total wavefield is known on sampling points that are located in each of the two configurations shown in Fig. 7.

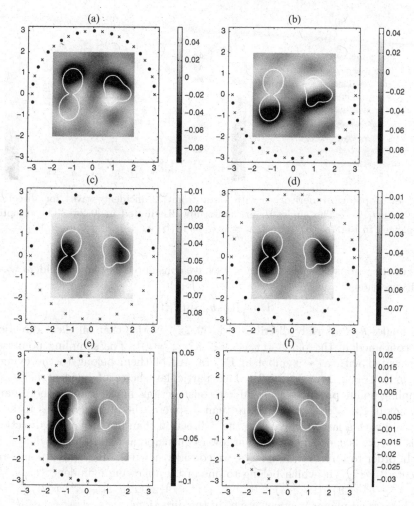

Fig. 6. Topological derivative when the incident fields are generated at the source points marked with '•' and the total field is measured at the receptors marked with '×'. $\lambda_e = 2.5$, $\lambda_i = 0.5$, $\alpha_i = 1$

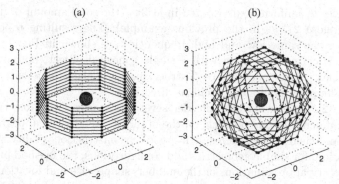

Fig. 7. Location of the observation points

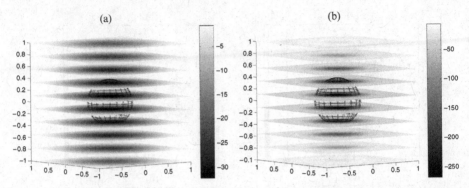

Fig. 8. Topological derivative with data from 22 incident waves for directions $(\cos\theta_j, \sin\theta_j, 0)^\top$. The angles θ_j are uniformly distributed in $[0, 2\pi)$. The receptors are plotted in Fig. 7a. (**a**) $\lambda_e = 2$, $\lambda_i = 1$, $\alpha_i = 1$, (**b**) $\lambda_e = 4$, $\lambda_i = 2$, $\alpha_i = 1$

For the reconstructions in Fig. 8, we have considered 21 planar incident waves with directions

$$\mathbf{d}_j := (\cos\theta_j, \sin\theta_j, 0)^\top$$

and angles θ_j uniformly distributed in $[0, 2\pi)$. Notice that all these directions are contained in the plane $\{(x, y, z) \in \mathbb{R}^3, \ z = 0\}$. The sampling points are represented with a '•' symbol in Fig. 7a. All of them belong to the cylinder $\{(x, y, z) \in \mathbb{R}^3, \ x^2 + y^2 = 9\}$. This particular choice of incident fields and sampling points produces acceptable reconstructions in the x and y coordinates, but not in the z coordinate. More than a sphere, the obstacle seems to be an ellipsoid with a larger semiaxis in the z-direction than in the x and y directions. This feature is specially stressed at low frequencies, as shown in Fig. 8a.

In order to improve the results, we consider now the observation points represented in Fig. 7b, which belong to one of the following cylinders, $\{(x, y, z) \in \mathbb{R}^3, \ x^2 + y^2 = 9\}$, $\{(x, y, z) \in \mathbb{R}^3, \ x^2 + z^2 = 9\}$, $\{(x, y, z) \in \mathbb{R}^3, \ y^2 + z^2 = 9\}$. We have tried planar incident waves in the directions

$$(\cos\theta_j, \sin\theta_j, 0)^\top, \quad (\cos\theta_j, 0, \sin\theta_j)^\top, \quad (0, \cos\theta_j, \sin\theta_j)^\top,$$

for seven angles θ_j uniformly distributed in $[0, 2\pi)$ (the total amount of incident waves is therefore 21, as in the previous example). The resulting topological derivatives for the same low and high frequencies are shown in Fig. 9. In this simple geometrical configuration with a single spherical object, both low and high frequencies provide accurate reconstructions when choosing a suitable distribution of sampling points and incident directions.

As in the 2D case, when we consider several obstacles and more complex shapes, the influence of one obstacle on the others as well as the appearance of illuminated and shadow regions result in poorer reconstruction quality.

In conclusion, for simple geometries and configurations, the method produces quite satisfactory approximations of the number, shape, size and location of the obstacles. If we consider more complicated problems, then, the method provides

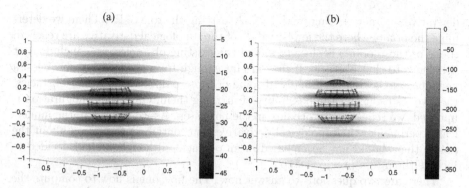

Fig. 9. Topological derivative with data from seven incident waves with directions $(\cos\theta_j, \sin\theta_j, 0)^\top$, seven incident waves with directions $(\cos\theta_j, 0, \sin\theta_j)^\top$ and seven incident waves with directions $(0, \cos\theta_j, \sin\theta_j)^\top$. The angles θ_j are uniformly distributed in $[0, 2\pi)$. The receptors are plotted in Fig. 7b. (**a**) $\lambda_e = 2$, $\lambda_i = 1$, $\alpha_i = 1$, (**b**) $\lambda_e = 4$, $\lambda_i = 2$, $\alpha_i = 1$

at least a good first guess for more sophisticated iterative methods which can circumvent the problems observed in the previous examples.

4 An Iterative Method Based on Topological Derivatives

There are several ways of improving the first guesses obtained by the method described in the previous section. One alternative is to make small deformations of the initial guess to reduce the cost functional, as done in [27]. However, this has the disadvantage of not allowing for topological changes in the obstacles, that is, the number of contours has to be known from the beginning. We are more interested in numerical methods with the ability of creating new obstacles, merging two or more contours or even destroying existing ones. A possibility is to use the approximation of the obstacle provided by the topological derivative combined with an optimization scheme based in level-set methods to improve the reconstruction, as done in [37] (see also [17,50]). Another option is to introduce the topological derivative in the Hamilton–Jacobi equation controlling the evolution of the level-sets, as proposed in [3].

In this section we will develop an efficient and fast numerical method based on the computation of successive topological derivatives. This approach was suggested in [21] for elasticity problems with Dirichlet and Neumann boundary conditions. We restrict to 2D experiments to reduce the computational cost in the tests, but the method applies without further difficulties (in principle) to 3D models. Again, the model problem will be the Helmholtz transmission problem, but the theory extends to other equations.

The idea is quite simple. We start by computing the topological derivative as in the previous section, that is, taking $\Omega_i = \emptyset$. Then, we look at the values of the topological derivative that fall below a certain negative threshold and create the initial guess Ω_1 of Ω_i. In the next iteration, we compute the topological derivative in a new setting, with $\mathbb{R}^2 \setminus \overline{\Omega}_1$ playing the role of $\mathbb{R}^2$. Then, we determine the points where the lowest values of the topological derivative are reached. If one of those points is close to any of the existing obstacles, we conclude that it belongs to that obstacle. If they are far enough from the existing obstacles, a new obstacle is created. Finally, if some of those points are near two obstacles, then we also consider the possibility of merging them. Once the configuration is updated, we repeat the procedure in the next step. As we will see in the numerical experiments, the topological derivative also contains information about the points that should not be included in an obstacle, although we had decided to include them in a previous step.

There are two questions to answer now. The first one is how to compute the topological derivative when $\Omega_i \neq \emptyset$ and the second is how to characterize the obstacles at each step in a suitable way for the numerical computations.

We will devote Sect. 5 to the first issue. The forward and the adjoint problems for $\Omega_i \neq \emptyset$ are solved by a fully discrete version of a Galerkin method with trigonometric polynomials applied to the system of boundary integral equations that is obtained when using Brakhage–Werner potentials.

Let us address the second question. Typically, boundary element discretizations are based on the hypothesis that one can describe the boundary Γ by a (smooth) parametrization of the boundary. We describe below a numerical method for the practical computation of such parametrization in the 2D case proposed in [4]. This is essential to determine the first guess Ω_1 knowing the topological derivative when $\Omega_i = \emptyset$, or in general, to construct the approximation Ω_{k+1} from the lowest values reached by the topological derivative when $\Omega_i = \Omega_k$.

First of all, by simple inspection we determine the number of components of Ω_i, that is, the number d such that $\Omega_i = \cup_{j=1}^{d} \Omega_{i,j}$. The idea now is to represent each obstacle $\Omega_{i,j}$ using a 1-periodic function $\mathbf{y}_j : \mathbb{R} \to \Gamma_j$ of the form

$$\mathbf{y}_j(t) := (c_1^j + r_j(t)\cos(2\pi t), \ c_2^j + r_j(t)\sin(2\pi t)),$$

where $r_j(t) : \mathbb{R} \to \mathbb{R}$ is also a 1-periodic function. The domains that admit this type of parametrization are usually said to be star-shaped.

The location of the obstacles and their centers (c_1^j, c_2^j) is decided by simple inspection. Then, we solve a least squares problem for each obstacle to find an approximation for the corresponding function $r_j(t)$ of the form

$$r_j(t) \approx a_0 + \sum_{k=1}^{K} a_k \cos(2\pi kt) + \sum_{k=1}^{K} b_k \sin(2\pi kt), \quad a_0, a_k, b_k \in \mathbb{R}.$$

The choice of K depends on the shape of the obstacle. Our numerical experiments suggest that, at low frequencies, the lowest values of the topological derivative are reached inside a circular or elliptical like region and therefore we simply choose $K = 1$. However, for higher frequencies, the lowest values of the topological derivative are located in annular regions enclosing the boundaries. Depending on the shape of the obstacle, one can decide that $K = 1$ is not enough and increase its value. For the numerical experiments that we will present below, we have taken in all cases $K = 1$ or $K = 2$ to compute initial guesses and values between $K = 1$ and $K = 5$ for the subsequent iterations. One could also look for 1-periodic spline functions $r_j(t)$, which are more flexible than trigonometric polynomials, to approximate difficult points if the original boundary Γ_j is smooth but not $\mathcal{C}^\infty$.

In three dimensions we can proceed in a similar way, looking for a parametrization in polar coordinates. Spherical harmonics play then the role of trigonometric polynomials.

We reproduce now the tests in Sect. 3 for the same geometries with one, two and three scatterers. After a few iterations, the description of the obstacles improves in a significant way.

First, we apply the iterative procedure in the configuration with a single scatterer. Figure 10 illustrates the performance of the method at low frequencies. Figure 10a is the equivalent of Fig. 2b but omitting the location of the observation points. The first guess, denoted by Ω_1, is superimposed on the representation of

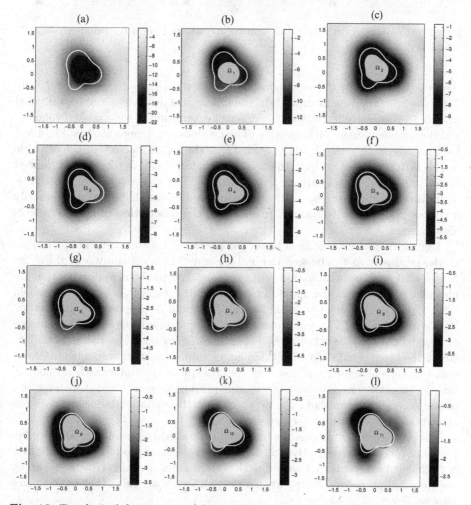

Fig. 10. Topological derivative and first 11 iterations. Parameter values are $\lambda_e = 2.5$, $\lambda_i = 0.5$ and $\alpha_i = 1$

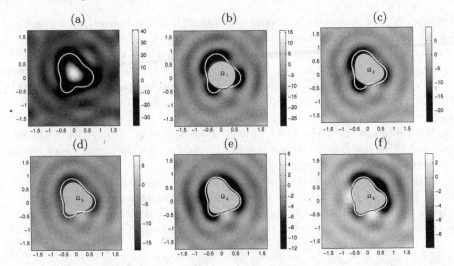

Fig. 11. Topological derivative and first five iterations. Parameter values are $\lambda_e = 5$, $\lambda_i = 1$ and $\alpha_i = 1$

the values attained by the topological derivative when $\Omega_i = \Omega_1$ given in Fig. 10b. After the first step, the magnitude of the topological derivative is almost divided by two. In successive iterations, the order of magnitude of the updated topological derivatives decays progressively. Their values are represented in Fig. 10c–l for the guesses $\Omega_i = \Omega_k, k = 1, \ldots, 11$. Ω_k is the reconstruction of the obstacle at the k-th step. Notice that the topological derivative at this step is defined in $\mathbb{R}^2 \setminus \overline{\Omega}_k$ and Ω_k appears as a solid region. The location of the obstacle is determined in the first guess. Its approximate size and shape are satisfactorily described after a few iterations. In 11 steps, the description of the obstacle is almost exact.

High frequencies allow for a faster reconstruction of the obstacle in this particular geometry, as can be seen in Fig. 11. Figure 11a is the equivalent of Fig. 2c. Notice that each approximation Ω_{k+1} contains the previous approximation Ω_k plus new points where the updated topological derivative falls below a threshold $-C_k$, $C_k > 0$. A first trial value for the threshold is proposed by inspection. We then determine Ω_{k+1} and update the topological derivative. In this sequence of approximations, we observe the appearance of regions close to the boundary where the topological derivative takes larger values (light colors). This indicates that our current approximation may contain spurious points which do not belong to the obstacle. We can observe this phenomenon in Fig. 11f. In practice, before accepting a threshold C_k, we check that the updated derivative does not show this anomaly. Otherwise, we increase C_k.

For more complex geometries involving several scatterers, the oscillatory behavior of topological derivatives at high frequencies may produce patterns which are difficult to interpret without a priori information. For this reason, we apply the iterative procedure at low frequencies, although we know that convergence may be slower.

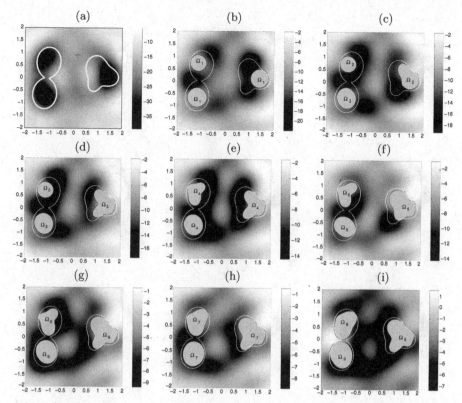

Fig. 12. Topological derivative and first eight iterations. Parameter values are $\lambda_e = 2.5$, $\lambda_i = 0.5$ and $\alpha_i = 1$

The results for the geometry with two scatterers, are presented in Fig. 12. The initial guess for the iterative scheme is computed using Fig. 4a, which seems to indicate the presence of three obstacles, one of them misplaced. After few iterations we find a good approximation of the real configuration. Figure 12 shows the first eight iterations. After three more iterations, the two obstacles on the left merge and the obstacle on the right is almost recovered. As expected, the biggest discrepancies are located in the region that is farther from the observation points and where the influence of one obstacle on the other is stronger.

Let us now consider the geometry with three scatterers. The first guess computed from Fig. 4c detects only the two largest obstacles, ignoring the small one. For computational simplicity, if a point is included in an obstacle, it remains inside for subsequent iterations. In the same way, once we have created an obstacle, we do not destroy it. It can only be merged with a neighboring one. Therefore, we are cautious when creating a new obstacle and, although Figs. 13b,c are pointing out the presence of the third obstacle, we have waited until the third iteration to create it. Again, a few iterations provide a good approximation to the true configuration.

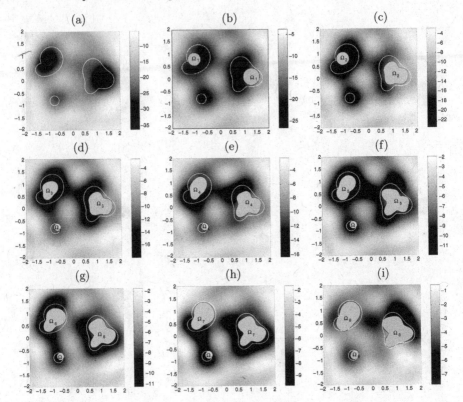

Fig. 13. Topological derivative and first eight iterations. Parameter values $\lambda_e = 2.5$, $\lambda_i = 0.5$ and $\alpha_i = 1$

The iterative procedure implemented in this section generates an increasing sequence of approximations $\Omega_k \subset \Omega_{k+1}$ (up to small deformations due to solving a problem in a least squares sense). This restriction comes from definition (5). If $\mathcal{R} = \mathbb{R}^2 \setminus \overline{\Omega}_k$, (5) defines the topological derivative only for points $\mathbf{x} \in \mathcal{R}$. The guess Ω_k can only be corrected by adding exterior points for which the topological derivative falls below a negative threshold. This restriction may be eliminated by extending the definition of topological derivative to $\mathbf{x} \notin \mathcal{R}$.

Notice that for the particular type of functional we consider here, we can remove balls both in $\mathcal{R} = \mathbb{R}^2 \setminus \overline{\Omega}$ and in Ω. We may define the topological derivative for $\mathbf{x} \in \mathbb{R}^2$ as follows

$$D_T(\mathbf{x}, \mathcal{R}) := \lim_{\varepsilon \to 0} \frac{\mathcal{J}_\varepsilon(\mathcal{R}) - \mathcal{J}(\mathcal{R})}{\mathcal{V}(\varepsilon)} \tag{24}$$

with $\mathcal{J}(\mathcal{R}) = J(\Omega)$ and $\mathcal{J}_\varepsilon(\mathcal{R})$ defined as $J(\Omega)$, but removing the ball $B_\varepsilon(\mathbf{x})$. In our case,

$$\mathcal{J}_\varepsilon(\mathcal{R}) = J_\varepsilon(\Omega) := \frac{1}{2} \int_{\Gamma_{meas}} |u_\varepsilon - u_{meas}|^2 dl,$$

where u_ε solves (1) with

$$\Omega_i = \Omega \setminus \overline{B}_\varepsilon(\mathbf{x}), \qquad \Omega_e = \mathbb{R}^2 \setminus \overline{\Omega}_i, \quad \text{if } \mathbf{x} \in \Omega,$$
$$\Omega_i = \Omega \cup B_\varepsilon(\mathbf{x}), \qquad \Omega_e = \mathbb{R}^2 \setminus \overline{\Omega}_i, \quad \text{if } \mathbf{x} \notin \Omega.$$

In this way, we might implement an iterative procedure in which $\Omega_{k+1} = \{\mathbf{x} \in \mathbb{R}^2 \,|\, D_T(\mathbf{x}, \mathbb{R}^2 \setminus \overline{\Omega}_k) < -C_{k+1}\}$. At each iteration, points can be added or removed, see [4] for a comparison of both strategies. This would allow to make holes inside an obstacle, for instance.

5 Numerical Solution of Forward and Adjoint Problems

Section 4 shows that efficient procedures to solve adjoint and forward problems are required if we want to implement topological derivative methods for inverse scattering problems. Techniques based in series expansions, finite elements or the method of moments are discussed in [18, 19, 37]. Inhomogeneous problems can be handled by coupled BEM–FEM methods, see [4]. For the models with constant parameters we are considering, integral methods are very competitive and can be easily adapted to 3D tests. We explain here the general setting for Helmholtz problems. The techniques can be extended to other boundary value problems by modifying the integral equations and fundamental solutions.

When $\Omega_i \neq \emptyset$, the forward and adjoint fields that appear in the formula of the topological derivative solve the Helmholtz transmission problem (1) and (12). We assume that Ω_i has d disjoint and connected components, that is, $\Omega_i = \cup_{j=1}^d \Omega_{i,j}$, with $\overline{\Omega}_{i,j} \cap \overline{\Omega}_{i,k}$ for $j \neq k$, and denote $\Gamma_j := \partial \Omega_{i,j}$. The theoretical and practical study of boundary element methods for this type of problems has attracted a lot of attention in the past decades. Different formulations using integral equations appear in [10, 13, 32, 55], with an emphasis on existence and uniqueness results. For a detailed description of the use of single and double layer potentials we refer to [10]. The recent papers [15, 45, 46, 48] deal with different boundary integral formulations and analyze a wide class of numerical methods.

In particular, a simple quadrature scheme based on the ideas of qualocation methods [6, 16, 51] that requires no implementation effort at all is studied in [15] (see also [48]). For the sake of brevity, we just describe in detail this method. This approach fails when either $-\widetilde{\lambda}_i^2$ or $-\lambda_e^2$ are Dirichlet eigenvalues of the Laplacian in any of the interior domains $\Omega_{i,j}$. In these cases, the scheme detects the proximity of the eigenvalues by means of a drastic increase in the condition number. At that point, the method must be changed. For instance, a formulation based on Brakhage–Werner potentials (a linear combination of a single and a double layer potential) can be used. The corresponding system of boundary integral equations includes hypersingular operators and the development and analysis of simple quadrature methods for these operators is still a challenging issue. We resort to full discretization using trigonometric polynomials, following [46]. This is the method we implement in our numerical computations to avoid possible eigenvalues. However, it is much more involved that the one we present here. For this reason, we will just give some remarks and references about it at the end of this section.

Let us describe the simple quadrature scheme for the forward and adjoint transmission problems in Sect. 4. First of all, we rewrite the transmission problem in terms of the scattered and the transmitted waves, which is more suitable for numerical purposes. The function

$$
v := \begin{cases} u - u_{inc}, & \text{in } \Omega_e, \\[2mm] u, & \text{in } \Omega_i = \cup_{j=1}^d \Omega_{i,j}, \end{cases}
$$

solves a problem equivalent to (1)

$$
\begin{cases} \Delta v + \lambda_e^2 v = 0, & \text{in } \Omega_e, \\[2mm] \Delta v + \widetilde{\lambda}_i^2 v = 0, & \text{in } \Omega_{i,j}, \quad j = 1, \dots, d, \\[2mm] v^- - v^+ = u_{inc}, & \text{on } \Gamma_j, \quad j = 1, \dots, d, \\[2mm] \alpha_i \partial_\mathbf{n} v^- - \partial_\mathbf{n} v^+ = \partial_\mathbf{n} u_{inc}, & \text{on } \Gamma_j, \quad j = 1, \dots, d, \\[2mm] \lim_{r \to \infty} r^{1/2}(\partial_r v - \imath \lambda_e v) = 0, \end{cases} \tag{25}
$$

with

$$
\widetilde{\lambda}_i := \lambda_i / \sqrt{\alpha_i}.
$$

The incident field appears in the transmission conditions instead of the radiation condition at infinity. We have switched to the standard notation in the boundary and finite element literature where $\mathbf{n}$ typically stands for the exterior normal vector pointing outside Ω_i. In the literature on topological derivatives $\mathbf{n}$ denotes the unit normal vector pointing inside the bounded domain Ω_i. We use this notation in Sects. 2, 6 and 7, so that the formulae agree with the ones in that literature. We only switch to the new notation in this section.

The adjoint field $\overline{p}$ solves

$$
\begin{cases} \Delta \overline{p} + \lambda_e^2 \overline{p} = \sum_{k=1}^M \overline{(u_{meas}(\mathbf{x}_k) - u(\mathbf{x}_k))}\, \delta_{\mathbf{x}_k}, & \text{in } \Omega_e, \\[3mm] \Delta \overline{p} + \widetilde{\lambda}_i^2 \overline{p} = 0, & \text{in } \Omega_{i,j}, \quad j = 1, \dots, d, \\[2mm] \overline{p}^- - \overline{p}^+ = 0, & \text{on } \Gamma_j, \quad j = 1, \dots, d, \\[2mm] \alpha_i \partial_\mathbf{n} \overline{p}^- - \partial_\mathbf{n} \overline{p}^+ = 0, & \text{on } \Gamma_j, \quad j = 1, \dots, d, \\[2mm] \lim_{r \to \infty} r^{1/2}(\partial_r \overline{p} - \imath \lambda_e \overline{p}) = 0. \end{cases}
$$

We decompose $\overline{p}$ as $\overline{p} = p_1 + p_2$ with

$$
p_1 := \begin{cases} \sum_{k=1}^N \overline{(u_{meas}(\mathbf{x}_k) - u(\mathbf{x}_k))}\, \phi_{\lambda_e}(\mathbf{x}, \mathbf{x}_k), & \text{in } \Omega_e, \\[3mm] 0, & \text{in } \Omega_{i,j}, \quad j = 1, \dots, d, \end{cases}
$$

ϕ_{λ_e} being the fundamental solution of the Helmholtz equation (see definition (21)). Recalling the results at the beginning of Sect. 3, the adjoint problem reduces to solving the following transmission problem for p_2:

$$
\begin{cases}
\Delta p_2 + \lambda_e^2 p_2 = 0, & \text{in } \Omega_e, \\
\Delta p_2 + \tilde{\lambda}_i^2 p_2 = 0, & \text{in } \Omega_{i,j}, \quad j = 1, \ldots, d, \\
p_2^- - p_2^+ = p_1^+, & \text{on } \Gamma_j, \quad j = 1, \ldots, d, \\
\alpha_i \partial_{\mathbf{n}} p_2^- - \partial_{\mathbf{n}} p_2^+ = \partial_{\mathbf{n}} p_1^+, & \text{on } \Gamma_j, \quad j = 1, \ldots, d, \\
\lim_{r \to \infty} r^{1/2} (\partial_r p_2 - \imath \lambda_e p_2) = 0.
\end{cases}
\tag{26}
$$

Notice that problems (25) and (26) are completely analogous. The only difference comes from the right hand sides. Therefore, to compute numerically the topological derivative one just has to assemble one matrix and solve two problems with different right hand sides. We denote the right hand sides by f_j and g_j. More precisely,

$$f_j = u_{inc}|_{\Gamma_j}, \quad g_j = \partial_{\mathbf{n}} u_{inc}|_{\Gamma_j}, \qquad \text{for the forward problem,}$$

$$f_j = p_1^+|_{\Gamma_j}, \quad g_j = \partial_{\mathbf{n}} p_1^+|_{\Gamma_j}, \qquad \text{for the adjoint problem.}$$

In case data for several incident waves are available, a couple of problems of the form (25) and (26) has to be solved for each of them. After discretization, all of them will share the same matrix and only the source terms change.

Let us assume that $\mathbf{y}_k : \mathbb{R} \to \Gamma_k$ is a smooth 1-periodic parametrization of Γ_k. We define the functions

$$V_{kj}^\lambda(s, t) := \phi_\lambda(\mathbf{y}_k(s), \mathbf{y}_j(t)),$$

$$J_{kj}^\lambda(s, t) := |\mathbf{y}_k'(s)| \partial_{\mathbf{n}_k(s)} \phi_\lambda(\mathbf{y}_k(s), \mathbf{y}_j(t)),$$

where $\mathbf{n}_k(s)$ is the outward normal vector of Γ_k at the point $\mathbf{y}_k(s)$. For the right hand sides, we set

$$f_k(s) := f_k(\mathbf{y}_k(s)), \qquad g_k(s) := |\mathbf{y}_k'(s)| g_k(\mathbf{y}_k(s)).$$

The method is as follows. We select a discretization parameter $n \in \mathbb{N}$ and define the meshes

$$t_p := p/n, \qquad t_{p+\varepsilon} := (p+\varepsilon)/n, \qquad p = 1, \ldots, n,$$

for $0 \neq \varepsilon \in (-1/2, 1/2)$. We take $\varepsilon \neq 0$ to avoid the singularity of the Hankel function and $\varepsilon \neq \pm 1/2$ for stability questions. Then, we solve the following linear problem: find

$$\varphi_{q,j}^i, \ \varphi_{q,j}^e \in \mathbb{C}, \qquad q = 1,\dots,n, \quad j = 1,\dots,d,$$

such that for $p = 1,\dots,n$ and $k = 1,\dots,d$,

$$\sum_{q=1}^n V_{kk}^{\tilde\lambda_i}(t_{p+\varepsilon}, t_q)\varphi_{q,k}^i - \sum_{j=1}^d \sum_{q=1}^n V_{kj}^{\lambda_e}(t_{p+\varepsilon}, t_q)\varphi_{q,j}^e$$

$$= f_k(t_{p+\varepsilon}), \qquad p = 1,\dots,n, \quad k = 1,\dots,d, \tag{27}$$

$$\alpha_i \left(\frac{1}{2} \varphi_{p,k}^i + \frac{1}{n} \sum_{q=1}^n J_{kk}^{\tilde\lambda_i}(t_p, t_q)\varphi_{q,k}^i \right) + \left(\frac{1}{2} \varphi_{p,k}^e + \frac{1}{n} \sum_{j=1}^d \sum_{q=1}^n J_{kj}^{\lambda_e}(t_p, t_q)\varphi_{q,k}^e \right)$$

$$= \frac{1}{n} g_k(t_p), \qquad p = 1,\dots,n, \quad k = 1,\dots,d. \tag{28}$$

Finally, the approximated solution to the transmission problem is computed as

$$u_n^\varepsilon(\mathbf{z}) := \begin{cases} \sum_{j=1}^d \sum_{q=1}^n \phi_{\lambda_e}(\mathbf{z}, \mathbf{y}_j(t_q))\varphi_{q,j}^e, & \mathbf{z} \in \Omega_e, \\ \sum_{q=1}^n \phi_{\tilde\lambda_i}(\mathbf{z}, \mathbf{y}_j(t_q))\varphi_{q,j}^i, & \mathbf{z} \in \Omega_{i,j}, \quad j = 1,\dots,d. \end{cases} \tag{29}$$

This method is a discrete version of a boundary integral formulation of the original Helmholtz transmission problem. We look for a function of the form

$$u = \begin{cases} \sum_{j=1}^d \mathcal{S}_j^{\lambda_e} \varphi_j^e, & \text{in } \Omega_e, \\ \mathcal{S}_j^{\tilde\lambda_i} \varphi_j^i & \text{in } \Omega_{i,j}, \quad j = 1,\dots,d, \end{cases} \tag{30}$$

where the single layer potentials $\mathcal{S}_j^\lambda$ are defined by

$$\mathcal{S}_j^\lambda \varphi := \int_0^1 \phi_\lambda(\,\cdot\,, \mathbf{y}_j(t))\varphi(t)dt, \tag{31}$$

and the densities φ_j^e, φ_j^i have to be determined. These densities are approximated then by Dirac delta distributions on the points $\mathbf{y}_j(t_q)$. The first set of equations (27) is obtained when testing the transmission conditions

$$u^- - u^+ = f_k, \qquad \text{on } \Gamma_k, \quad k = 1,\dots,d,$$

at the observation points $\mathbf{y}_k(t_{p+\varepsilon})$, (or equivalently, we are testing the equations with Dirac delta distributions on the points $\mathbf{y}_k(t_{p+\varepsilon})$). The second set of equations (28) is equivalent to the application of the classical Nyström method for the second transmission conditions

$$\alpha\partial_{\mathbf{n}}u^- - \partial_{\mathbf{n}}u^+ = g_k, \qquad \text{on } \Gamma_k, \qquad k = 1, \dots, d,$$

which in terms of the densities are integral equations of the second kind. For further details we refer to [15].

For any $0 \neq \varepsilon \in (-1/2, 1/2)$, we have a first order method. For the particular choices $\varepsilon = \pm 1/6$, the method has quadratic convergence order, that is,

$$|u(\mathbf{z}) - u_n^{\pm 1/6}(\mathbf{z})| \leq C_{\mathbf{z}}(1/n)^2$$

where $C_{\mathbf{z}} > 0$ is a constant that depends on $\mathbf{z}$ but not on n. Moreover, if we keep the $n \times d$ equations (28) unchanged and we replace

$$V_{kj}^\lambda(t_{p+\varepsilon}, t_q), \qquad (\text{for } \lambda = \lambda_e \text{ and } \lambda = \widetilde{\lambda}_i)$$

in (27) by the linear combination

$$\frac{5}{6}\left(V_{kj}^\lambda(t_{p-1/6}, t_q) + V_{kj}^\lambda(t_{p+1/6}, t_q)\right) + \frac{1}{6}\left(V_{kj}^\lambda(t_{p-5/6}, t_q) + V_{kj}^\lambda(t_{p+5/6}, t_q)\right)$$

as well as the value $f_k(t_{p+\varepsilon})$ at the right hand side by

$$\frac{5}{6}\left(f_k(t_{p-1/6}) + f_k(t_{p+1/6})\right) + \frac{1}{6}\left(f_k(t_{p-5/6}) + f_k(t_{p+5/6})\right)$$

we obtain a new linear system of equations to compute the unknown values $\varphi_{q,j}^i$, $\varphi_{q,j}^e \in \mathbb{C}$, $q = 1, \dots, n$, $j = 1, \dots, d$. If we define the discrete solution u_n^ε as in (29), then the method has cubic convergence order,

$$|u(\mathbf{z}) - u_n^\varepsilon(\mathbf{z})| \leq C_{\mathbf{z}}(1/n)^3.$$

This strategy was found using the ideas of qualocation methods, a variation of collocation methods that improves the order of convergence (see [6, 16, 51]).

This method is specially well suited for our purposes because moving the boundary requires almost no additional computational effort in recomputing the elements in the matrix of the linear system of equations. As we have already mentioned, this technique cannot be applied in the singular cases associated with interior Dirichlet eigenvalues of the Laplace operator. To overcome this problem, one can represent the solution of the transmission problem as

$$u := \begin{cases} \displaystyle\sum_{j=1}^d (\mathcal{S}_j^{\lambda_e} - \imath\eta\mathcal{D}_j^{\lambda_e})\varphi_j^e, & \text{in } \Omega_e, \\[2mm] (\mathcal{S}_j^{\widetilde{\lambda}_i} - \imath\eta\mathcal{D}_j^{\widetilde{\lambda}_i})\varphi_j^i, & \text{in } \Omega_{i,j}, \quad j = 1, \dots, d, \end{cases} \tag{32}$$

where $\eta > 0$ is a fixed parameter, $\mathcal{S}_j^\lambda$ are the single layer potentials that were introduced in (31) and the double layer potentials $\mathcal{D}_j^\lambda$ are defined by

$$\mathcal{D}_j^\lambda := \int_0^1 |\mathbf{y}_j'(t)|\, \partial_{\mathbf{n}_j(t)}\phi_\lambda(\,\cdot\,, \mathbf{y}_j(t))\varphi(t)dt.$$

Linear combinations of single and double layer potentials of the form $\mathcal{S} - \imath\eta\mathcal{D}$ with $\eta > 0$ are usually called mixed or Brakhage–Werner potentials. The system of integral equations found by imposing the transmission conditions at the interfaces Γ_k is more complicated than the system obtained for the simpler representation (30). Now, we deal with hypersingular operators. The equivalence of this formulation to the original problem was shown in [46]. We also refer to that work for a complete analysis of convergence of a wide class of Petrov–Galerkin methods. In the numerical tests presented in Sect. 4, we used a fully discrete Galerkin method with trigonometric polynomials that has superalgebraic convergence order in terms of the discretization parameter, avoiding therefore the occurrence of eigenvalues.

In 3D both integral representations (30) and (32) remain valid using the corresponding fundamental solution in 3D given by (21). The resultant systems of integral equations can be solved by using Petrov–Galerkin methods, as indicated in [45, 46]. For the implementation of fully discrete Galerkin methods with trigonometric polynomials we refer to [40]. Discretizations of Petrov–Galerkin methods for spline functions are studied in [45].

6 Explicit Expressions for the Topological Derivatives

Different strategies for the practical computation of topological derivatives are found in the literature. For instance, in [18], a useful connection with the shape derivative is given. Integral expressions in terms of Green functions of Helmholtz equations are exploited in [24]. Our procedure to obtain analytical expressions for topological derivatives combines both of them. First, we calculate the shape derivative of the functionals involved. Then, we perform asymptotic expansions of the solutions of Helmholtz equations with vanishing holes. Such expansions are performed directly on the partial differential equation formulation.

6.1 Shape Derivative

The shape derivative of the functional $\mathcal{J}(\mathcal{R}) = J(\Omega_i)$ in the direction of a smooth vector field $\mathbf{V}(\mathbf{x})$ is defined as

$$DJ(\Omega_i) \cdot \mathbf{V} := \frac{d}{d\tau} J(\phi_\tau(\Omega_i))\Big|_{\tau=0},$$

where $\phi_\tau(\mathbf{x}) := \mathbf{x} + \tau\mathbf{V}(\mathbf{x})$, $\mathbf{x} \in \mathbb{R}^2$ and $\mathbf{V}$ represents the direction of change of the domain.

For practical computations, a more precise formula in terms of solutions of two auxiliary direct and adjoint problems is used. When $\mathbf{V}$ satisfies $\mathbf{V} = V_n\mathbf{n}$ on $\partial\Omega_i$ and $\mathbf{V} = 0$ out of a small neighborhood of $\partial\Omega_i$, we have the following result.

Theorem 6.1. *The shape derivative of the functional J defined in* (3) *is given by*

$$DJ(\Omega_i) \cdot \mathbf{V} = \mathrm{Re}\left[\int_\Gamma (-\alpha_i \nabla u^- \nabla \overline{p}^- + \lambda_i^2 u^- \overline{p}^- + 2\alpha_i \partial_\mathbf{n} u^- \partial_\mathbf{n} \overline{p}^-) V_n \, dl\right.$$

$$\left. - \int_\Gamma (-\nabla u^+ \nabla \overline{p}^+ + \lambda_e^2 u^+ \overline{p}^+ + 2\partial_\mathbf{n} u^+ \partial_\mathbf{n} \overline{p}^+) V_n \, dl\right], \quad (33)$$

where $V_n = \mathbf{V} \cdot \mathbf{n}$, u *is the solution of the forward problem and* p *is the continuous solution of an adjoint problem. For particular choices of the boundary conditions at the interface between the medium and the obstacles,* (33) *takes the following simpler forms:*

- *General transmission problem (penetrable obstacle):*

$$DJ(\Omega_i) \cdot \mathbf{V} = \mathrm{Re}\left[\int_\Gamma (\alpha_i(1 - \alpha_i)\partial_\mathbf{n} u^- \partial_\mathbf{n} \overline{p}^- + (1 - \alpha_i)\partial_t u^- \partial_t \overline{p}^-) V_n \, dl\right.$$

$$\left. + \int_\Gamma (\lambda_i^2 - \lambda_e^2) u \overline{p} V_n \, dl\right]. \quad (34)$$

  where u *solves* (1) *and* p *solves* (12).
- *Transmission problem with* $\alpha_i = 1$

$$DJ(\Omega_i) \cdot \mathbf{V} = \mathrm{Re}\left[\int_\Gamma (\lambda_i^2 - \lambda_e^2) u \overline{p} V_n \, dl\right], \quad (35)$$

  where u *and* p *solve the forward and adjoint transmission problems* (1) *and* (12) *with* $\alpha_i = 1$.
- *Neumann problem (sound hard obstacle):*

$$DJ(\Omega_i) \cdot \mathbf{V} = \mathrm{Re}\left[\int_\Gamma (\nabla u \nabla \overline{p} - \lambda_e^2 u \overline{p}) V_n \, dl\right], \quad (36)$$

  where u *and* p *solve the exterior homogeneous Neumann problems* (2) *and* (15).

This result is independent of the value of $\mathbf{V}$ outside Γ. Notice that (33) is rewritten in terms only of the interior values of the solutions of the forward and adjoint problems by using the transmission boundary conditions and splitting the gradients in tangential and normal components:

$$\nabla u = (\nabla u \cdot \mathbf{t})\mathbf{t} + (\nabla u \cdot \mathbf{n})\mathbf{n} = \partial_t u \, \mathbf{t} + \partial_\mathbf{n} u \, \mathbf{n}, \quad \text{on } \Gamma.$$

The continuity of u across Γ implies the continuity of the tangential derivatives: $\partial_t u^- = \partial_t u^+$. When $\alpha_i = 1$, the normal derivatives are also continuous and the continuity of the gradients at the interface follows.

Formula (33) would extend to other boundary conditions by simply changing the conditions at the interface between the obstacle and the surrounding medium in the forward and adjoint problems.

Here, we only prove (33) for the generalized transmission problem. Related results are found in [4] for transmission problems in heterogeneous media in [17, 37] for the transmission problem with α_i and in [19] for the Neumann problem.

Proof. We adapt the strategy introduced in [19] to derive (36) for the Neumann problem.

The idea is the following. First, we give a variational formulation of the boundary value problem in a bounded domain using transparent boundary conditions. Next, we deform the domain Ω_i along the vector field $\mathbf{V}$ and compute the shape functional in the deformed domains. Then, we differentiate the transformed functionals with respect to the control parameter τ. The resulting expression involves the derivative of the solutions u_τ in the deformed domains. The computation of this derivative is avoided establishing a relationship with the derivatives of modified functionals, in which a free state can be selected to eliminate $\frac{du_\tau}{d\tau}$. This adjoint state solves the so-called adjoint problem. Integrating by parts the resulting expression involving adjoint solutions, we find the desired formula for the shape derivative. The proof is organized in four steps.

Step 1: Variational formulation of the forward problem. First, we replace the exterior problem (1) by an equivalent boundary value problem posed in a bounded domain. Let us introduce a circle Γ_R which encloses the obstacles as in Fig. 1. The Dirichlet-to-Neumann (also called Steklov–Poincaré) operator associates to any Dirichlet data on Γ_R the normal derivative of the solution of the exterior Dirichlet problem:

$$L : H^{1/2}(\Gamma_R) \longrightarrow H^{-1/2}(\Gamma_R)$$

$$f \longmapsto \partial_{\mathbf{n}} w$$

where $w \in H^1_{loc}(\mathbb{R}^2 \setminus \overline{B}_R)$, $B_R := B(\mathbf{0}, R)$, is the unique solution of

$$\begin{cases} \Delta w + \lambda_e^2 w = 0, & \text{in } \mathbb{R}^2 \setminus \overline{B}_R, \\ w = f, & \text{on } \Gamma_R, \\ \lim_{r \to \infty} r^{1/2}(\partial_r w - \imath \lambda_e w) = 0. \end{cases}$$

$H^1_{loc}(\mathbb{R}^2 \setminus \overline{B}_R)$ denotes the usual Sobolev space and $H^{1/2}(\Gamma_R)$ and $H^{-1/2}(\Gamma_R)$ are the standard trace spaces. A detailed representation of the Dirichlet-to-Neumann map in terms of Hankel functions is given in [19, 28]. Instead of the exterior transmission problem (1) one can study an equivalent boundary value problem in B_R with a non-reflecting boundary condition on Γ_R:

$$\begin{cases} \Delta u + \lambda_e^2 u = 0, & \text{in } \Omega'_e := B_R \setminus \overline{\Omega}_i, \\[1mm] \alpha_i \Delta u + \lambda_i^2 u = 0, & \text{in } \Omega_i, \\[1mm] u^- - u^+ = 0, & \text{on } \Gamma, \\[1mm] \alpha_i \partial_{\mathbf{n}} u^- - \partial_{\mathbf{n}} u^+ = 0, & \text{on } \Gamma, \\[1mm] \partial_{\mathbf{n}}(u - u_{inc}) = L(u - u_{inc}), & \text{on } \Gamma_R. \end{cases} \qquad (37)$$

The solution u of (37) also solves the variational equation

$$\begin{cases} u \in H^1(B_R), \\[1mm] b(\Omega_i; u, v) = \ell(v), & \forall v \in H^1(B_R), \end{cases} \qquad (38)$$

where

$$b(\Omega_i; u, v) := \int_{\Omega'_e} (\nabla u \nabla \overline{v} - \lambda_e^2 u \overline{v}) d\mathbf{z} + \int_{\Omega_i} (\alpha_i \nabla u \nabla \overline{v} - \lambda_i^2 u \overline{v}) d\mathbf{z}$$

$$- \int_{\Gamma_R} L u \, \overline{v} \, dl, \qquad \forall u, v \in H^1(B_R),$$

$$\ell(v) := \int_{\Gamma_R} (\partial_{\mathbf{n}} u_{inc} - L u_{inc}) \, \overline{v} \, dl, \qquad \forall v \in H^1(B_R).$$

Step 2: Transformed functionals. Since $\mathbf{V}$ decreases rapidly to zero away from Γ, we have $\phi_\tau(\Gamma_R) = \Gamma_R$ and $\phi_\tau(\Gamma_{meas}) = \Gamma_{meas}$. Set

$$\Omega_{i,\tau} := \phi_\tau(\Omega_i), \qquad \Omega'_{e,\tau} := \phi_\tau(\Omega'_e) = B_R \setminus \overline{\Omega}_{i,\tau}.$$

The variational reformulations (38) of the original Helmholtz problem in the deformed domains take the form

$$\begin{cases} u_\tau \in H^1(B_R), \\[1mm] b(\Omega_{i,\tau}; u_\tau, v) = \ell(v), & \forall v \in H^1(B_R), \end{cases} \qquad (39)$$

with

$$b(\Omega_{i,\tau}; u, v) := \int_{\Omega'_{e,\tau}} (\nabla_{\mathbf{z}_\tau} u \nabla_{\mathbf{z}_\tau} \overline{v} - \lambda_e^2 u \overline{v}) d\mathbf{z}_\tau + \int_{\Omega_{i,\tau}} (\alpha_i \nabla_{\mathbf{z}_\tau} u \nabla_{\mathbf{z}_\tau} \overline{v} - \lambda_i^2 u \overline{v}) d\mathbf{z}_\tau$$

$$- \int_{\Gamma_R} L u \, \overline{v} \, dl, \qquad \forall u, v \in H^1(B_R).$$

The cost functional in the transformed domains is

$$J(\Omega_{i,\tau}) = \frac{1}{2} \int_{\Gamma_{meas}} |u_\tau - u_{meas}|^2 dl, \tag{40}$$

where u_τ solves (39). Solving (39) with $\tau = 0$ we recover the solution $u = u_0$ of (38).

Step 3: Adjoint states. Differentiating (40) with respect to τ we obtain

$$\frac{d}{d\tau} J(\Omega_{i,\tau}) \Big|_{\tau=0} = \mathrm{Re} \left[\int_{\Gamma_{meas}} \overline{(u - u_{meas})} \dot{u} \, dl \right], \tag{41}$$

where $\dot{u} = \frac{d}{d\tau} u_\tau |_{\tau=0}$. To avoid computing $\dot{u}$ we introduce an adjoint problem. Let us define a family of modified functionals

$$\mathcal{L}(\Omega_{i,\tau}; v, p) = J(\Omega_{i,\tau}) + \mathrm{Re}[b(\Omega_{i,\tau}; v, p) - \ell(p)], \qquad \forall v, p \in H^1(B_R),$$

where p plays the role of a Lagrangian multiplier. Setting $v = u_\tau$, we obtain

$$\mathcal{L}(\Omega_{i,\tau}; u_\tau, p) = J(\Omega_{i,\tau}), \qquad \forall p \in H^1(B_R). \tag{42}$$

The shape derivative of J in the direction $\mathbf{V}$ is then

$$DJ(\Omega_i) \cdot \mathbf{V} = \frac{d}{d\tau} \mathcal{L}(\Omega_{i,\tau}; u_\tau, p) \Big|_{\tau=0} = \mathrm{Re} \left[\frac{d}{d\tau} b(\Omega_{i,\tau}; u, p) \Big|_{\tau=0} \right]$$

$$+ \mathrm{Re} \left[b(\Omega_i; \dot{u}, p) \right] + \mathrm{Re} \left[\int_{\Gamma_{meas}} \overline{(u - u_{meas})} \dot{u} \, dl \right], \tag{43}$$

thanks to (41) and (42). This is true for any $p \in H^1(\Omega_R)$. If p solves

$$\begin{cases} p \in H^1(B_R), \\ b(\Omega_i; v, p) = \int_{\Gamma_{meas}} \overline{(u_{meas} - u)} \, v \, dl, \qquad \forall v \in H^1(B_R), \end{cases} \tag{44}$$

the derivative $\dot{u}$ is not needed and the shape derivative $DJ(\Omega) \cdot \mathbf{V}$ is given by the first term in (43). We select the solution p of (44) as an adjoint state. The adjoint problem (44) is equivalent to

$$\begin{cases} \overline{p} \in H^1(B_R), \\ b(\Omega_i; \overline{p}, v) = \int_{\Gamma_{meas}} (u_{meas} - u) \, \overline{v} \, dl, \qquad \forall v \in H^1(B_R), \end{cases} \tag{45}$$

which has the same structure as the direct problem (38) and p is then a solution of (12). A property of the Dirichlet-to-Neumann operator is essential in this argument: its adjoint operator satisfies $L^*(\overline{u}) = \overline{L(u)}$.

Step 4: Shape derivative. Let us take p to be the adjoint state and compute the first term in (43). The change of variables from the deformed to the original variables is governed by the identities

$$\nabla_{\mathbf{z}_\tau} u = F_\tau^{-\top} \nabla u, \qquad d\mathbf{z}_\tau = \det F_\tau \, d\mathbf{z},$$

and

$$\left. \frac{d}{d\tau} F_\tau \right|_{\tau=0} = \nabla \mathbf{V}, \qquad \left. \frac{d}{d\tau} (\nabla_{\mathbf{z}_\tau} u) \right|_{\tau=0} = -\nabla \mathbf{V}^\top \nabla u, \qquad \left. \frac{d}{d\tau} d\mathbf{z}_\tau \right|_{\tau=0} = \nabla \cdot \mathbf{V} d\mathbf{z},$$

where $F_\tau := \nabla \phi_\tau = I + \tau \nabla \mathbf{V}$ is the deformation gradient. For $u, p \in H^1(B_R)$,

$$\left. \frac{d}{d\tau} b(\Omega_{i,\tau}; u, p) \right|_{\tau=0} =$$

$$\int_{\Omega_e'} (\nabla u \cdot \nabla \overline{p} - \lambda_e^2 u \overline{p}) \nabla \cdot \mathbf{V} \, d\mathbf{z} - \int_{\Omega_e'} ((\nabla \mathbf{V} + \nabla \mathbf{V}^\top) \nabla u) \cdot \nabla \overline{p} \, d\mathbf{z} +$$

$$\int_{\Omega_i} (\alpha_i \nabla u \cdot \nabla \overline{p} - \lambda_i^2 u \overline{p}) \nabla \cdot \mathbf{V} \, d\mathbf{z} - \int_{\Omega_i} \alpha_i ((\nabla \mathbf{V} + \nabla \mathbf{V}^\top) \nabla u) \cdot \nabla \overline{p} \, d\mathbf{z}. \quad (46)$$

We may further simplify this expression integrating by parts:

$$\int_{\Omega_i} (\alpha_i \nabla u \cdot \nabla \overline{p} - \lambda_i^2 u \overline{p}) \nabla \cdot \mathbf{V} \, d\mathbf{z} =$$

$$\lambda_i^2 \int_\Gamma u^- \overline{p}^- \mathbf{V} \cdot \mathbf{n} \, dl - \alpha_i \int_\Gamma (\nabla u^- : \nabla \overline{p}^-) \mathbf{V} \cdot \mathbf{n} \, dl +$$

$$\lambda_i^2 \int_{\Omega_i} \left(\frac{\partial u}{\partial x_\ell} \overline{p} + u \frac{\partial \overline{p}}{\partial x_\ell} \right) v_\ell \, d\mathbf{z} - \alpha_i \int_{\Omega_i} \left(\frac{\partial^2 u}{\partial x_j \partial x_\ell} \frac{\partial \overline{p}}{\partial x_j} v_\ell + \frac{\partial^2 \overline{p}}{\partial x_j \partial x_\ell} \frac{\partial u}{\partial x_j} v_\ell \right) d\mathbf{z} \quad (47)$$

and

$$-\int_{\Omega_i} ((\nabla \mathbf{V} + \nabla \mathbf{V}^\top) \nabla u) \cdot \nabla \overline{p} \, d\mathbf{z} =$$

$$\int_{\Omega_i} \left(\frac{\partial^2 u}{\partial x_j^2} \frac{\partial \overline{p}}{\partial x_\ell} v_\ell + \frac{\partial^2 \overline{p}}{\partial x_j^2} \frac{\partial u}{\partial x_\ell} v_\ell \right) d\mathbf{z} + \int_{\Omega_i} \left(\frac{\partial^2 u}{\partial x_j \partial x_\ell} \frac{\partial \overline{p}}{\partial x_j} v_\ell + \frac{\partial^2 \overline{p}}{\partial x_j \partial x_\ell} \frac{\partial u}{\partial x_j} v_\ell \right) d\mathbf{z}$$

$$+ \int_\Gamma \left(\frac{\partial u^-}{\partial x_j} \frac{\partial \overline{p}^-}{\partial x_\ell} v_\ell n_j + \frac{\partial \overline{p}^-}{\partial x_j} \frac{\partial u^-}{\partial x_\ell} v_\ell n_j \right) d\mathbf{z}, \quad (48)$$

where $\mathbf{V} = (v_1, v_2)^\top$, $\mathbf{n} = (n_1, n_2)^\top$ and summation over repeated indexes is understood. Recall that $\mathbf{n}$ points inside Ω_i.

Let us now add the contributions from (47) and from (48) multiplied by α_i. Part of the integrals over Ω_i cancel and the rest vanishes

$$\int_{\Omega_i} (\alpha_i \Delta u + \lambda_i^2 u) \nabla \overline{p} \cdot \mathbf{V} \, d\mathbf{z} + \int_{\Omega_i} (\alpha_i \Delta \overline{p} + \lambda_i^2 \overline{p}) \nabla u \cdot \mathbf{V} \, d\mathbf{z} = 0$$

because u and $\overline{p}$ satisfy Helmholtz equations in Ω_i. The sum of the integrals over Γ is

$$-\alpha_i \int_\Gamma (\nabla u^- \cdot \nabla \overline{p}^-) V_n \, dl + \lambda_i^2 \int_\Gamma u^- \overline{p}^- V_n \, dl + 2\alpha_i \int_\Gamma (\nabla u^- \cdot \mathbf{n})(\nabla \overline{p}^- \cdot \mathbf{n}) V_n \, dl.$$

using $\mathbf{V} = V_n \, \mathbf{n}$.

We reproduce the same computations for Ω_e'. All the integrals over Γ_R are identically zero, since $\mathbf{V} = 0$ on Γ_R. Recall that $\mathbf{n}$ is now the outward normal vector. Integrating by parts we obtain

$$\int_{\Omega_e'} (\nabla u \cdot \nabla \overline{p} - \lambda_e^2 u \overline{p}) \nabla \cdot \mathbf{V} \, d\mathbf{z} - \int_{\Omega_e'} ((\nabla \mathbf{V} + \nabla \mathbf{V}^\top) \nabla u) \cdot \nabla \overline{p} \, d\mathbf{z}$$

$$= \int_{\Omega_e'} (\Delta u + \lambda_e^2 u) \nabla \overline{p} \cdot \mathbf{V} \, d\mathbf{z} + \int_{\Omega_e'} (\Delta \overline{p} + \lambda_e^2 \overline{p}) \nabla u \cdot \mathbf{V} \, d\mathbf{z} +$$

$$\int_\Gamma (\nabla u^+ \nabla \overline{p}^+) V_n \, dl - \lambda_e^2 \int_\Gamma u^+ \overline{p}^+ V_n \, dl - 2 \int_\Gamma (\nabla u^+ \cdot \mathbf{n})(\nabla \overline{p}^+ \cdot \mathbf{n}) V_n \, dl. \quad (49)$$

Notice that both u and $\overline{p}$ solve the Helmholtz equations

$$\Delta u + \lambda_e^2 u = 0, \qquad \Delta \overline{p} + \lambda_e^2 \overline{p} = \overline{(u_{meas} - u)}\, \delta_{\Gamma_{meas}}, \qquad \text{in } \Omega_e'.$$

Due to these identities and the fact that $\mathbf{V} = 0$ on Γ_{meas}, the two integrals over Ω_e' vanish.

Adding all the integrals on Γ generated integrating by parts on Ω_i and Ω_e', we finally find (33).   □

6.2 Proof of Theorems 2.1 and 2.2

Theorem 2.1.- Transmission problem without holes. Let us first calculate the topological derivative of the cost functional (3) for the transmission problem when $\mathcal{R} = \mathbb{R}^2$. Knowing the formula of the shape derivative for the transmission problem, we use the identities (6) and (34) to obtain

$$D_T(\mathbf{x}, \mathbb{R}^2) = \lim_{\varepsilon \to 0} \frac{-1}{\mathcal{V}'(\varepsilon)} \, \text{Re} \Bigg[\int_{\Gamma_\varepsilon} \bigg(\alpha_i (1 - \alpha_i) \partial_\mathbf{n} u_\varepsilon^- \, \partial_\mathbf{n} \overline{p}_\varepsilon^- \qquad (50)$$

$$+ (1 - \alpha_i) \partial_\mathbf{t} u_\varepsilon^- \, \partial_\mathbf{t} \overline{p}_\varepsilon^- \bigg) \, dl + \int_{\Gamma_\varepsilon} (\lambda_i^2 - \lambda_e^2) u_\varepsilon \overline{p}_\varepsilon \, dl \Bigg].$$

Recall that V_n is constant and negative. Here, u_ε and p_ε solve the forward and adjoint problems when $\Omega_i = B_\varepsilon(\mathbf{x})$ and $\Gamma = \Gamma_\varepsilon = \partial B_\varepsilon(\mathbf{x})$. A value for this limit is computed performing an asymptotic expansion of these solutions and their gradients at Γ_ε as in [4].

The asymptotic behavior of u_ε and p_ε is obtained expressing them as corrections of u and p

$$u_\varepsilon(\mathbf{z}) = u(\mathbf{z}) \chi_{\mathbb{R}^2 \setminus \overline{B}_\varepsilon}(\mathbf{z}) + v_\varepsilon(\mathbf{z}), \qquad p_\varepsilon(\mathbf{z}) = p(\mathbf{z}) \chi_{\mathbb{R}^2 \setminus \overline{B}_\varepsilon}(\mathbf{z}) + q_\varepsilon(\mathbf{z}),$$

and expanding the remainders in powers of ε.

Let us denote by B the unit ball. Changing variables $\boldsymbol{\xi} := (\mathbf{z} - \mathbf{x})/\varepsilon$, the correction $v_\varepsilon(\boldsymbol{\xi})$ satisfies

$$
\begin{cases}
\Delta_{\boldsymbol{\xi}} v_\varepsilon + \varepsilon^2 \lambda_e^2 v_\varepsilon = 0, & \text{in } \mathbb{R}^2 \setminus \overline{B}, \\[2mm]
\alpha_i \Delta_{\boldsymbol{\xi}} v_\varepsilon + \varepsilon^2 \lambda_i^2 v_\varepsilon = 0, & \text{in } B, \\[2mm]
v_\varepsilon^- - v_\varepsilon^+ = u(\mathbf{z}) = u(\mathbf{x}) + \varepsilon \boldsymbol{\xi} \cdot \nabla u(\mathbf{x}) + O(\varepsilon^2), & \text{on } \Gamma, \\[2mm]
\alpha_i \mathbf{n}(\boldsymbol{\xi}) \cdot \nabla_{\boldsymbol{\xi}} v_\varepsilon^- - \mathbf{n}(\boldsymbol{\xi}) \cdot \nabla_{\boldsymbol{\xi}} v_\varepsilon^+ = \varepsilon \mathbf{n}(\boldsymbol{\xi}) \cdot \nabla u(\mathbf{x}) + O(\varepsilon^2), & \text{on } \Gamma, \\[2mm]
\lim_{r \to \infty} r^{1/2} \left(\partial_r v_\varepsilon - \imath \varepsilon \lambda_e v_\varepsilon \right) = 0, & r = |\boldsymbol{\xi}|.
\end{cases}
\tag{51}
$$

Let us expand now $v_\varepsilon(\boldsymbol{\xi})$ in powers of ε : $v_\varepsilon(\boldsymbol{\xi}) = v^{(1)}(\boldsymbol{\xi}) + \varepsilon v^{(2)}(\boldsymbol{\xi}) + O(\varepsilon^2)$. The leading terms of the expansion solve:

$$
\begin{cases}
\Delta_{\boldsymbol{\xi}} v^{(1)} = 0, & \text{in } \mathbb{R}^2 \setminus \overline{B} \text{ and } B, \\[2mm]
v^{(1)-} - v^{(1)+} = u(\mathbf{x}), & \text{on } \Gamma, \\[2mm]
\alpha_i \mathbf{n}(\boldsymbol{\xi}) \cdot \nabla_{\boldsymbol{\xi}} v^{(1)-} - \mathbf{n}(\boldsymbol{\xi}) \cdot \nabla_{\boldsymbol{\xi}} v^{(1)+} = 0, & \text{on } \Gamma, \\[2mm]
\lim_{r \to \infty} r^{1/2} \partial_r v^{(1)} = 0,
\end{cases}
\tag{52}
$$

and

$$
\begin{cases}
\Delta_{\boldsymbol{\xi}} v^{(2)} = 0, & \text{in } \mathbb{R}^2 \setminus \overline{B} \text{ and } B, \\[2mm]
v^{(2)-} - v^{(2)+} = \boldsymbol{\xi} \cdot \nabla u(\mathbf{x}), & \text{on } \Gamma, \\[2mm]
\alpha_i \mathbf{n}(\boldsymbol{\xi}) \cdot \nabla_{\boldsymbol{\xi}} v^{(2)-} - \mathbf{n}(\boldsymbol{\xi}) \cdot \nabla_{\boldsymbol{\xi}} v^{(2)+} = \mathbf{n}(\boldsymbol{\xi}) \cdot \nabla u(\mathbf{x}), & \text{on } \Gamma, \\[2mm]
\lim_{r \to \infty} r^{1/2} \left(\partial_r v^{(2)} - \imath \lambda_e v^{(1)} \right) = 0.
\end{cases}
\tag{53}
$$

By inspection, $v^{(1)}(\boldsymbol{\xi}) = u(\mathbf{x}) \chi_B(\boldsymbol{\xi})$. The second term, $v^{(2)}(\boldsymbol{\xi})$, can be found working in polar coordinates:

$$
v^{(2)}(\boldsymbol{\xi}) = \nabla u(\mathbf{x}) \cdot \left(\frac{2}{1+\alpha_i} \boldsymbol{\xi} \chi_B(\boldsymbol{\xi}) + \frac{1-\alpha_i}{1+\alpha_i} \frac{\boldsymbol{\xi}}{|\boldsymbol{\xi}|^2} \chi_{\mathbb{R}^2 \setminus \overline{B}}(\boldsymbol{\xi}) \right) = \nabla u(\mathbf{x}) \cdot g(\boldsymbol{\xi}).
$$

Thus,

$$
u_\varepsilon(\mathbf{z}) = u(\mathbf{z}) \chi_{\mathbb{R}^2 \setminus \overline{B}_\varepsilon}(\mathbf{z}) + u(\mathbf{x}) \chi_{B_\varepsilon}(\mathbf{z}) + \varepsilon \nabla u(\mathbf{x}) \cdot g(\boldsymbol{\xi}) + O(\varepsilon^2), \quad \mathbf{z} \in \Gamma_\varepsilon.
$$

Performing a similar expansion for $p_\varepsilon(\mathbf{z})$, we get

$$
p_\varepsilon(\mathbf{z}) = p(\mathbf{z}) \chi_{\mathbb{R}^2 \setminus \overline{B}_\varepsilon}(\mathbf{z}) + p(\mathbf{x}) \chi_{B_\varepsilon}(\mathbf{z}) + \varepsilon \nabla p(\mathbf{x}) \cdot g(\boldsymbol{\xi}) + O(\varepsilon^2), \quad \mathbf{z} \in \Gamma_\varepsilon.
$$

This implies that

$$u_\varepsilon(\mathbf{z}) \to u(\mathbf{x}), \qquad p_\varepsilon(\mathbf{z}) \to p(\mathbf{x}), \qquad \text{as } \varepsilon \to 0,$$

uniformly when $|\mathbf{z} - \mathbf{x}| = \varepsilon$. Expanding the derivatives we find

$$\frac{\partial u_\varepsilon^-}{\partial z_j}(\mathbf{z}) = \varepsilon \frac{\partial v^{(2)-}}{\partial z_j}(\boldsymbol{\xi}) + O(\varepsilon) = \frac{\partial v^{(2)-}}{\partial \xi_j}(\boldsymbol{\xi}) + O(\varepsilon) = \frac{2}{1+\alpha_i} \frac{\partial u}{\partial z_j}(\mathbf{x}) + O(\varepsilon),$$

and a similar identity for p_ε. Therefore,

$$\nabla u_\varepsilon^-(\mathbf{z}) \to \frac{2}{1+\alpha_i} \nabla u(\mathbf{x}), \qquad \nabla p_\varepsilon^-(\mathbf{z}) \to \frac{2}{1+\alpha_i} \nabla p(\mathbf{x}), \qquad \text{as } \varepsilon \to 0.$$

Let us take limits in the three integrals appearing in (50). For the integral of the normal components, we write the unit normal as a function of the angle $(n_1, n_2) = (\cos\theta, \sin\theta)$. Then,

$$\int_{\Gamma_\varepsilon} (\nabla u_\varepsilon^- \cdot \mathbf{n})(\nabla \overline{p}_\varepsilon \cdot \mathbf{n}) dl_z \approx \varepsilon \frac{4}{(1+\alpha_i)^2} \frac{\partial u_\varepsilon}{\partial z_j}(\mathbf{x}) \frac{\partial \overline{p}_\varepsilon}{\partial z_\ell}(\mathbf{x}) \int_0^{2\pi} n_j n_\ell d\theta$$

$$= \varepsilon \frac{4\pi}{(1+\alpha_i)^2} \nabla u(\mathbf{x}) \nabla \overline{p}(\mathbf{x}),$$

plus a remainder of higher order in ε, where we have used that $\int_0^{2\pi} n_1 n_2\, d\theta = 0$ and $\int_0^{2\pi} n_j^2\, d\theta = \pi$ for $j = 1, 2$. To compute the integral of the tangential components, we write the tangent vector as $(t_1, t_2) = (-\sin\theta, \cos\theta)$. Exactly the same value is found. Finally, the third integral is

$$\int_{\Gamma_\varepsilon} u_\varepsilon \overline{p}_\varepsilon dl_z \approx 2\varepsilon\pi\, u(\mathbf{x})\overline{p}(\mathbf{x}).$$

Taking into account that $\mathcal{V}'(\varepsilon) = -2\pi\varepsilon$, (7) follows. Formula (8) is a particular case with $\alpha_i = 1$.

Theorem 2.1.- Neumann problem without holes. The analytical expression of the topological derivative for Neumann problems (9) follows by combining (6) and (36). Then, we perform an asymptotic expansion of u_ε and p_ε, which now solve Neumann problems. We find that

$$u_\varepsilon(\mathbf{z}) = u(\mathbf{z}) + \varepsilon \nabla u(\mathbf{x}) \cdot \frac{\boldsymbol{\xi}}{|\boldsymbol{\xi}|^2} + O(\varepsilon^2).$$

Thus, $u_\varepsilon(\mathbf{z}) = u(\mathbf{x}) + O(\varepsilon)$ and

$$\frac{\partial u_\varepsilon}{\partial z_j}(\mathbf{z}) = \frac{\partial u}{\partial z_j}(\mathbf{x}) + \nabla u(\mathbf{x}) \cdot \frac{\partial}{\partial \xi_j} \frac{\boldsymbol{\xi}}{|\boldsymbol{\xi}|^2} + O(\varepsilon)$$

as $\varepsilon \to 0$, uniformly when $|\mathbf{z} - \mathbf{x}| = \varepsilon$. Similar expressions hold for p_ε and ∇p_ε. Computing the limit, we obtain (9).

Theorem 2.2.- Transmission problem with holes. Let us calculate now the topological derivative of the cost functional (3) in a domain with a hole $\mathcal{R} = \mathbb{R}^2 \setminus \overline{\Omega}$, Ω being an open bounded set, not necessarily connected. Formula (11) follows by slightly modifying the procedure we have used to compute (7) in $\mathbb{R}^2$ for the transmission problem.

Now, $u_\varepsilon(\mathbf{z}) = u(\mathbf{z})\chi_{\mathbb{R}^2 \setminus \overline{B}}(\mathbf{z}) + v_\varepsilon(\mathbf{z})$, where u solves (1) with $\Omega_i = \Omega$. Changing variables, $v_\varepsilon(\boldsymbol{\xi})$ satisfies

$$
\begin{cases}
\Delta_{\boldsymbol{\xi}} v_\varepsilon + \varepsilon^2 \lambda_e^2 v_\varepsilon = 0, & \text{in } \mathbb{R}^2 \setminus (\overline{B} \cup \overline{\Omega}_\varepsilon), \\[2mm]
\alpha_i \Delta_{\boldsymbol{\xi}} v_\varepsilon + \varepsilon^2 \lambda_i^2 v_\varepsilon = 0, & \text{in } B \cup \Omega_\varepsilon, \\[2mm]
v_\varepsilon^- - v_\varepsilon^+ = u(\mathbf{z}) = u(\mathbf{x}) + \varepsilon\,\boldsymbol{\xi} \cdot \nabla u(\mathbf{x}) + O(\varepsilon^2), & \text{on } \Gamma, \\[2mm]
\alpha_i \mathbf{n}(\boldsymbol{\xi}) \cdot \nabla_{\boldsymbol{\xi}} v_\varepsilon^- - \mathbf{n}(\boldsymbol{\xi}) \cdot \nabla_{\boldsymbol{\xi}} v_\varepsilon^+ = \varepsilon \mathbf{n}(\boldsymbol{\xi}) \cdot \nabla u(\mathbf{x}) + O(\varepsilon^2), & \text{on } \Gamma, \\[2mm]
v_\varepsilon^- - v_\varepsilon^+ = 0, & \text{on } \partial\Omega_\varepsilon, \\[2mm]
\alpha_i \mathbf{n}(\boldsymbol{\xi}) \cdot \nabla_{\boldsymbol{\xi}} v_\varepsilon^- - \mathbf{n}(\boldsymbol{\xi}) \cdot \nabla_{\boldsymbol{\xi}} v_\varepsilon^+ = 0, & \text{on } \partial\Omega_\varepsilon, \\[2mm]
\lim_{r \to \infty} r^{1/2} \left(\partial_r v_\varepsilon - \imath\varepsilon\lambda_e v_\varepsilon\right) = 0, & r = |\boldsymbol{\xi}|,
\end{cases}
\tag{54}
$$

with $\Omega_\varepsilon := (\Omega - \mathbf{x})/\varepsilon$.

Expanding $v_\varepsilon(\boldsymbol{\xi})$ in powers of ε, the leading terms $v^{(1)}(\boldsymbol{\xi})$, $v^{(2)}(\boldsymbol{\xi})$ solve again (52) and (53), respectively, and (11) follows. We just have to check that the presence of Ω_ε does not provide corrections to the orders zero and one. Let us consider the boundary value problems:

$$
\begin{cases}
\Delta_{\boldsymbol{\xi}} v^{(1)} = 0, & \text{in } \mathbb{R}^2 \setminus (\overline{B} \cup \overline{\Omega}_\varepsilon) \text{ and } B \cup \Omega_\varepsilon, \\[2mm]
v^{(1)-} - v^{(1)+} = u(\mathbf{x}), & \text{on } \Gamma, \\[2mm]
\alpha_i \mathbf{n}(\boldsymbol{\xi}) \cdot \nabla_{\boldsymbol{\xi}} v^{(1)-} - \mathbf{n}(\boldsymbol{\xi}) \cdot \nabla_{\boldsymbol{\xi}} v^{(1)+} = 0, & \text{on } \Gamma, \\[2mm]
v^{(1)-} - v^{(1)+} = 0, & \text{on } \partial\Omega_\varepsilon, \\[2mm]
\alpha_i \mathbf{n}(\boldsymbol{\xi}) \cdot \nabla_{\boldsymbol{\xi}} v^{(1)-} - \mathbf{n}(\boldsymbol{\xi}) \cdot \nabla_{\boldsymbol{\xi}} v^{(1)+} = 0, & \text{on } \partial\Omega_\varepsilon, \\[2mm]
\lim_{r \to \infty} r^{1/2} \partial_r v^{(1)} = 0,
\end{cases}
\tag{55}
$$

and

$$
\begin{cases}
\Delta_{\boldsymbol{\xi}} v^{(2)} = 0, & \text{in } \mathbb{R}^2 \setminus (\overline{B} \cup \overline{\Omega}_\varepsilon) \text{ and } B \cup \Omega_\varepsilon, \\[2mm]
v^{(2)-} - v^{(2)+} = \boldsymbol{\xi} \cdot \nabla u(\mathbf{x}), & \text{on } \Gamma, \\[2mm]
\alpha_i \mathbf{n}(\boldsymbol{\xi}) \cdot \nabla_{\boldsymbol{\xi}} v^{(2)-} - \mathbf{n}(\boldsymbol{\xi}) \cdot \nabla_{\boldsymbol{\xi}} v^{(2)+} = \mathbf{n}(\boldsymbol{\xi}) \cdot \nabla u(\mathbf{x}), & \text{on } \Gamma, \\[2mm]
v^{(2)-} - v^{(2)+} = 0, & \text{on } \partial \Omega_\varepsilon, \\[2mm]
\alpha_i \mathbf{n}(\boldsymbol{\xi}) \cdot \nabla_{\boldsymbol{\xi}} v_\varepsilon^{(2)-} - \mathbf{n}(\boldsymbol{\xi}) \cdot \nabla_{\boldsymbol{\xi}} v^{(2)+} = 0, & \text{on } \partial \Omega_\varepsilon, \\[2mm]
\lim_{r \to \infty} r^{1/2} \left(\partial_r v^{(2)} - \imath \lambda_e v^{(1)} \right) = 0.
\end{cases} \tag{56}
$$

Then, $v^{(1)}(\boldsymbol{\xi}) = u(\mathbf{x}) \chi_B(\boldsymbol{\xi})$ is still a solution of (55). Since $\alpha_i \neq 0$,

$$
v^{(2)}(\boldsymbol{\xi}) = \nabla u(\mathbf{x}) \cdot \left(\frac{2}{1+\alpha_i} \boldsymbol{\xi} \chi_B(\boldsymbol{\xi}) + \frac{1-\alpha_i}{1+\alpha_i} \frac{\boldsymbol{\xi}}{|\boldsymbol{\xi}|^2} \chi_{\mathbb{R}^2 \setminus \overline{B}}(\boldsymbol{\xi}) \right),
$$

solves (56) with an error of order ε^2. Thus, the correction coming from Ω_ε appears at orders higher than ε.

A similar argument works for the Neumann problem.

7 Sounding Solids by Elastic Waves

The ideas and techniques developed in the previous sections for scattering problems involving waves governed by Helmholtz equations can be extended to more general situations. In this section, we consider the problem of detecting solids buried in an elastic medium by means of acoustic waves. We only describe the procedure to compute explicit formulae for the topological derivatives in this new setting. Once these formulae are known, the numerical approximation procedures described in Sects. 3–5 apply using the fundamental solutions and integral operators of elasticity theory.

7.1 The Forward and Inverse Scattering Problems

The general setting is similar to that in Sect. 2. The obstacle $\Omega_i \subset \mathbb{R}^2$ is an open bounded set with smooth boundary $\Gamma := \partial \Omega_i$ but has no assumed connectivity. The scattering pattern is measured at Γ_{meas} far enough from the scatterers, as in Fig. 1.

Consider the 2D elastodynamic problem in a solid, where the elastic fields depend only on the coordinates z_1, z_2 and time. The Navier equations of motion are given by

$$
\rho \frac{\partial^2 U_j}{\partial t^2} - c_{j\alpha m\beta} \frac{\partial^2 U_m}{\partial z_\alpha \partial z_\beta} = 0,
$$

where $c_{j\alpha m\beta}$ are the stiffness tensor components, ρ is the mass density and $\mathbf{z}$ and t are the 2D position vector and time, respectively. Subscripts range from 1 to 2. We adopt the convention that a repeated index is summed over its range.

Regardless of particular material symmetries, the elastic moduli satisfy the following symmetry restrictions:

$$c_{j\alpha m\beta} = c_{m\beta j\alpha} = c_{\alpha j\beta m},$$

together with the coercivity inequality:

$$\xi_{j\alpha} c_{j\alpha m\beta} \overline{\xi}_{m\beta} \geq C \sum_{j,\alpha=1}^{2} |\xi_{j\alpha}|^2$$

for some constant $C > 0$ and every complex $\xi_{j\alpha}$ such that $\xi_{j\alpha} = \xi_{\alpha j}$. The elastic constants $c_{j\alpha m\beta}$ for a cubic crystal are:

$$c_{j\alpha m\beta} = c_{12}\delta_{j\alpha}\delta_{m\beta} + c_{44}(\delta_{jm}\delta_{\alpha\beta} + \delta_{j\beta}\delta_{\alpha m}) - H\delta_{j\alpha m\beta} \tag{57}$$

where H is the anisotropy factor $H = 2c_{44} + c_{12} - c_{11}$. δ_{jm} stands for the Kronecker delta. In the isotropic case, $c_{12} = \lambda$, $c_{44} = \mu$ and $c_{11} = \lambda + 2\mu$.

We are interested in time harmonic solutions, in which the elastic displacement field can be written in the form $U_j(\mathbf{z}, t) = \mathrm{Re}\,[u_j(\mathbf{z})e^{-\imath\omega t}]$ for a given frequency $\omega > 0$. Therefore, the components $u_j, j = 1, 2$ solve

$$\rho\omega^2 u_j + c_{j\alpha m\beta}\frac{\partial^2 u_m}{\partial z_\alpha \partial z_\beta} = 0. \tag{58}$$

In the scattering process, a time harmonic vector plane wave of unit amplitude $\mathbf{u}_{inc}$ illuminates an object with section Ω_i embedded in an infinite elastic medium. This process generates two additional time harmonic fields: the scattered vector field $\mathbf{u}_{sc}$, defined outside the obstacle and propagating outwards, together with the transmitted vector field $\mathbf{u}_{tr}$, defined inside the obstacle and trapped in it.

Dropping the time harmonic dependence, which is implicit in (58), the incident waves have the form $\mathbf{u}_{inc}(\mathbf{z}) = \mathbf{u}_0 e^{\imath\mathbf{k}\cdot\mathbf{z}}$, with $\mathbf{z} = (z_1, z_2)$ and $\mathbf{k} = (k_1, k_2) = k\mathbf{d}$, where k is the wave number and the unitary vector $\mathbf{d}$ is the direction of propagation. The wave vector $\mathbf{k}$, the frequency ω and $\mathbf{u}_0$ are related by:

$$(\rho\omega^2\delta_{jm} - c_{j\alpha m\beta}k_\alpha k_\beta)u_{0,m} = 0, \qquad j = 1, 2. \tag{59}$$

A non-zero solution exists only if the determinant of this matrix is zero:

$$\det\left[\left(\rho\omega^2\delta_{jm} - c_{j\alpha m\beta}k_\alpha k_\beta\right)_{jm}\right] = 0. \tag{60}$$

Note that the frequency ω depends on the wave vector $\mathbf{k}$ through the dispersion relation (60). Substituting each of its two roots $\omega^2 = \omega_j^2(\mathbf{k})$ in system (59), we find the directions of displacement $\mathbf{u}_0$ in these waves (directions of polarization).

The equations are homogeneous, they do not determine their magnitude. We normalize them by imposing $|\mathbf{u_0}| = 1$. We may find two different directions of polarization for the same wave vector, which are perpendicular to each other since the matrix tensor $(c_{j\alpha m\beta}k_\alpha k_\beta)_{jm}$ is symmetrical. They are eigenvectors for different eigenvalues. Neither of these directions is in general purely longitudinal or purely transverse to $\mathbf{k}$.

For isotropic materials the dependence of ω on $\mathbf{k}$ simplifies to direct proportionality to its magnitude k. The branch $\omega = c_p k_p$, with $c_p = \sqrt{(\lambda + 2\mu)/\rho}$, corresponds to longitudinally polarized waves. The branch $\omega = c_s k_s$, with $c_s = \sqrt{\mu/\rho}$, corresponds to transversely polarized waves. In this case, incident waves take the form $\mathbf{u}_{inc}(\mathbf{z}) = \mathbf{u_0}e^{ik\mathbf{d}\cdot\mathbf{z}}$ where the wave number k is either k_p or k_s and the direction of polarization $\mathbf{u_0}$ is either $\mathbf{d}$ or is orthogonal to $\mathbf{d}$ (obtained by rotating $\mathbf{d}$ anticlockwise $\pi/2$).

The interaction between the scatterer, the medium and the incident radiation is described by the following transmission model where the total wavefield

$$\mathbf{u} = \begin{cases} \mathbf{u}_{inc} + \mathbf{u}_{sc}, & \text{in } \Omega_e := \mathbb{R}^2 \setminus \overline{\Omega}_i, \\ \mathbf{u}_{tr}, & \text{in } \Omega_i, \end{cases}$$

satisfies:

$$\begin{cases} \rho^i\omega^2 u_j + c^i_{j\alpha m\beta}\dfrac{\partial^2 u_m}{\partial z_\alpha \partial z_\beta} = 0, & j = 1,2, & \text{in } \Omega_i, \\[2mm] \rho^e\omega^2 u_j + c^e_{j\alpha m\beta}\dfrac{\partial^2 u_m}{\partial z_\alpha \partial z_\beta} = 0, & j = 1,2, & \text{in } \Omega_e, \\[2mm] \mathbf{u}^- - \mathbf{u}^+ = 0, & & \text{on } \Gamma := \partial\Omega_i, \\[2mm] c^i_{j\alpha m\beta}\dfrac{\partial u_m^-}{\partial z_\beta}n_\alpha - c^e_{j\alpha m\beta}\dfrac{\partial u_m^+}{\partial z_\beta}n_\alpha = 0, & j = 1,2, & \text{on } \Gamma, \end{cases} \tag{61}$$

supplemented by radiation conditions at infinity on the scattered field $\mathbf{u}_{sc} = \mathbf{u} - \mathbf{u}_{inc}$ implying that only outgoing waves are allowed. The dependence of densities and elastic constants on the domain is indicated by the superscripts i and e. We have kept the notation introduced in Sect. 2: $\mathbf{n}$ is the outward unit normal vector pointing outside Ω_e, and $\mathbf{u}^+$ and $\mathbf{u}^-$ denote the limits of $\mathbf{u}$ from the exterior and interior of Ω_i.

The radiation condition is a decay condition at infinity which selects the correct fundamental solution and implies uniqueness. For isotropic materials, any solution of (58) can be decomposed as $\mathbf{u} = \mathbf{u}^p + \mathbf{u}^s$ into a compressional part $\mathbf{u}^p$ (or longitudinal wave, satisfying $\mathrm{rot}(\mathbf{u}^p) = 0$) and a shear part $\mathbf{u}^s$ (or transverse wave, satisfying $\mathrm{div}(\mathbf{u}^s) = 0$). They satisfy the vector Helmholtz equation with wave numbers k_p and k_s, respectively. The Kupradze radiation condition for isotropic materials reads:

$$\lim_{r\to\infty} r^{1/2}\left(\partial_r(\mathbf{u}^c - \mathbf{u}^c_{inc}) - ik_c(\mathbf{u}^c - \mathbf{u}^c_{inc})\right) = 0, \qquad c = p,s \quad r = |\mathbf{x}|. \tag{62}$$

Solutions satisfying this condition are called radiating solutions of the isotropic 2D Navier equations, see the reviews [22, 35] and references therein. Radiation conditions for anisotropic materials are discussed in [22].

For rigid obstacles, there is no transmitted field. Then, the transmission problem is replaced by a Neumann problem:

$$
\begin{cases}
\rho^e \omega^2 u_j + c^e_{j\alpha m\beta} \dfrac{\partial^2 u_m}{\partial z_\alpha \partial z_\beta} = 0, & j = 1, 2, & \text{in } \Omega_e, \\[4mm]
c^e_{j\alpha m\beta} \dfrac{\partial u_m}{\partial z_\beta} n_\alpha = 0, & j = 1, 2, & \text{on } \Gamma,
\end{cases}
\tag{63}
$$

with a 'sound hard' boundary condition at the obstacle boundary plus the radiation condition on $\mathbf{u} - \mathbf{u}_{inc}$ at infinity. For simplicity, in this section we will work only with the Neumann problem and we will adopt the simpler notation $\rho \equiv \rho^e$ and $c_{j\alpha m\beta} \equiv c^e_{j\alpha m\beta}$. Similar, but more involved computations can be carried out for the transmission problem (see [5]).

The forward problem consists in computing the solution $\mathbf{u}$ of (63) at the measurement curve Γ_{meas} knowing the scatterers Ω_i and the incident wave $\mathbf{u}_{inc}$. Existence and uniqueness results for problems (61) and (63) can be found in [1, 7, 20, 25, 35, 42, 43].

As before, the inverse problem consists in finding the shape and structure of the obstacle Ω_i such that the solution of the forward Neumann problem (63) equals the measured values $\mathbf{u}_{meas}$ at the receptors, knowing the incident field. A less restrictive formulation is given by the following constrained optimization problem: minimize

$$
J(\Omega_i) := \frac{1}{2} \int_{\Gamma_{meas}} |\mathbf{u} - \mathbf{u}_{meas}|^2
\tag{64}
$$

where $\mathbf{u}$ is the solution of the forward Neumann problem (63) and $\mathbf{u}_{meas}$ the total field measured on Γ_{meas}.

7.2 Shape Derivatives

The computation of the topological derivative of the shape functional (64) follows the same steps as in Sect. 6. Let us first compute the shape derivative of (64). Then, we will use its expression to find an explicit formula for the topological derivative.

Theorem 7.1. *Keeping the notations of the previous sections, the shape derivative of the functional J defined in (64) is given by*

$$
DJ(\Omega) \cdot \mathbf{V} = \mathrm{Re} \left[\int_\Gamma \left(c_{j\alpha m\beta} \frac{\partial u_j}{\partial z_\alpha} \frac{\partial \overline{p}_m}{\partial z_\beta} - \rho \omega^2 \mathbf{u} \overline{\mathbf{p}} \right) V_n \, dl \right],
\tag{65}
$$

where $\mathbf{p}$ is a continuous solution of the adjoint problem

$$
\begin{cases}
\rho\omega^2 p_j + c_{j\alpha m\beta}\dfrac{\partial^2 p_m}{\partial z_\alpha \partial z_\beta} = (u_{meas} - u)_j \delta_{\Gamma_{meas}}, & j = 1, 2, & in\ \Omega_e, \\[3mm]
c_{j\alpha m\beta}\dfrac{\partial p_m}{\partial z_\beta} n_\alpha = 0, & j = 1, 2, & on\ \Gamma,
\end{cases}
\tag{66}
$$

together with a conjugated radiation condition at infinity for **p**.

Proof. As in the Helmholtz case, we prove the result in four steps, following the ideas in [5].

Step 1: Variational formulation of the forward problem in a bounded domain. First, we replace the unbounded Neumann problem by an equivalent problem posed in a finite domain using Steklov–Poincaré operators as non-reflecting boundary conditions in the artificial boundary.

We introduce a circle Γ_R which encloses the obstacles and define the Steklov–Poincaré operator on Γ_R, see Fig. 1. This operator associates to any Dirichlet data on Γ_R the elastic conormal derivative of the solution of the exterior Dirichlet problem in $\mathbb{R}^2 \setminus \overline{B}_R$, $B_R := B(\mathbf{0}, R)$:

$$
L : (H^{1/2}(\Gamma_R))^2 \longrightarrow (H^{-1/2}(\Gamma_R))^2
$$

$$
\mathbf{g} \longmapsto \partial_{\mathbf{n}}\mathbf{w} = \left(c_{j\alpha m\beta}\frac{\partial w_m}{\partial z_\beta} n_\alpha \right)_j
$$

where $\mathbf{w} \in (H^1(B_R))^2$ is the unique radiating solution of

$$
\begin{cases}
\rho\omega^2 u_j + c_{j\alpha m\beta}\dfrac{\partial^2 u_m}{\partial z_\alpha \partial z_\beta} = 0, & j = 1, 2 & in\ \mathbb{R}^2 \setminus \overline{B}_R, \\[3mm]
\mathbf{w} = \mathbf{g}, & & on\ \Gamma_R.
\end{cases}
$$

For a detailed representation of the Steklov–Poincaré operator in terms of integral kernels, we refer to [20] and references therein.

Then, we may replace (63) by an equivalent boundary value problem in B_R with a transparent boundary condition on the artificial boundary Γ_R:

$$
\begin{cases}
\rho\omega^2 u_j + c_{j\alpha m\beta}\dfrac{\partial^2 u_m}{\partial z_\alpha \partial z_\beta} = 0, & j = 1, 2, & in\ \Omega'_e := B_R \setminus \overline{\Omega}_i, \\[3mm]
c_{j\alpha m\beta}\dfrac{\partial u_m}{\partial z_\beta} n_\alpha = 0, & j = 1, 2, & on\ \Gamma, \\[3mm]
c_{j\alpha m\beta}\dfrac{\partial (u - u_{inc})_m}{\partial z_\beta} n_\alpha = L(\mathbf{u} - \mathbf{u}_{inc})_j, & j = 1, 2, & on\ \Gamma_R.
\end{cases}
\tag{67}
$$

Problem (67) admits the variational formulation

$$
\begin{cases}
\mathbf{u} \in (H^1(B_R))^2, \\[2mm]
b(\Omega_i; \mathbf{u}, \mathbf{v}) = \ell(\mathbf{v}), & \forall \mathbf{v} \in (H^1(B_R))^2,
\end{cases}
\tag{68}
$$

where

$$b(\Omega_i; \mathbf{u}, \mathbf{v}) := \int_{\Omega'_e} \left(c_{j\alpha m\beta} \frac{\partial u_j}{\partial z_\alpha} \frac{\partial \overline{v}_m}{\partial z_\beta} - \rho\omega^2 \mathbf{u}\overline{\mathbf{v}} \right) d\mathbf{z} - \int_{\Gamma_R} L\mathbf{u}\,\overline{\mathbf{v}}\,dl,$$

$$\forall \mathbf{u}, \mathbf{v} \in (H^1(B_R))^2,$$

$$\ell(\mathbf{v}) := \int_{\Gamma_R} \left(c_{j\alpha m\beta} \frac{\partial u_{inc,m}}{\partial z_\beta} n_\alpha \overline{v}_j - L\mathbf{u}_{inc}\,\overline{\mathbf{v}} \right) dl, \qquad \forall \mathbf{v} \in (H^1(B_R))^2.$$

Step 2: Transformed problems. Let u_τ be the solution of the variational formulations in the transformed domains:

$$\begin{cases} \mathbf{u}_\tau \in (H^1(B_R))^2, \\[2mm] b(\Omega_{i,\tau}; \mathbf{u}_\tau, \mathbf{v}) = \ell(\mathbf{v}), \qquad \forall v \in (H^1(B_R))^2, \end{cases} \tag{69}$$

with

$$b(\Omega_{i,\tau}; \mathbf{u}, \mathbf{v}) := \int_{\Omega'_{e,\tau}} \left(c_{j\alpha m\beta} \frac{\partial u_j}{\partial z_{\tau,\alpha}} \frac{\partial \overline{v}_m}{\partial z_{\tau,\beta}} - \rho\omega^2 \mathbf{u}\overline{\mathbf{v}} \right) d\mathbf{z}_\tau$$

$$- \int_{\Gamma_R} L\mathbf{u}\,\overline{\mathbf{v}}\,dl_\tau, \qquad \forall \mathbf{u}, \mathbf{v} \in (H^1(B_R))^2.$$

When $\tau = 0$, $\mathbf{u}_0 = \mathbf{u}$ is the solution to problem (68). The transformed functionals are

$$J(\Omega_{i,\tau}) = \frac{1}{2} \int_{\Gamma_{meas}} |\mathbf{u}_\tau - \mathbf{u}_{meas}|^2 \, dl. \tag{70}$$

Step 3: Adjoint problem. Differentiating (70) with respect to τ we obtain

$$\frac{d}{d\tau} J(\Omega_{i,\tau})\bigg|_{\tau=0} = \mathrm{Re}\left[\int_{\Gamma_{meas}} (\overline{\mathbf{u} - \mathbf{u}_{meas}})\dot{\mathbf{u}} \right], \tag{71}$$

where $\dot{\mathbf{u}} = \frac{d}{d\tau}\mathbf{u}_\tau\big|_{\tau=0}$. We define the Lagrangian functional

$$\mathcal{L}(\Omega_{i,\tau}; \mathbf{v}, \mathbf{p}) = J(\Omega_{i,\tau}) + \mathrm{Re}[b(\Omega_{i,\tau}; \mathbf{v}, \mathbf{p}) - \ell(\mathbf{p})], \qquad \forall \mathbf{v}, \mathbf{p} \in (H^1(B_R))^2.$$

When $\mathbf{v} = \mathbf{u}_\tau$ solves (69),

$$\mathcal{L}(\Omega_{i,\tau}; \mathbf{u}_\tau, \mathbf{p}) = J(\Omega_{i,\tau}), \qquad \forall \mathbf{p} \in (H^1(B_R))^2. \tag{72}$$

Using (71)–(72), the shape derivative of J in the direction $\mathbf{V}$ is

$$DJ(\Omega_i) \cdot \mathbf{V} = \frac{d}{d\tau} \mathcal{L}(\Omega_{i,\tau}; \mathbf{u}_\tau, \mathbf{p})\bigg|_{\tau=0} = \mathrm{Re}\left[\frac{d}{d\tau} b(\Omega_{i,\tau}; \mathbf{u}, \mathbf{p})\bigg|_{\tau=0} \right]$$

$$+ \mathrm{Re}\left[b(\Omega_i; \dot{\mathbf{u}}, \mathbf{p}) \right] + \mathrm{Re}\left[\int_{\Gamma_{meas}} (\overline{\mathbf{u} - \mathbf{u}_{meas}})\dot{\mathbf{u}} \right]. \tag{73}$$

Choosing $\mathbf{p}$ to be the solution of

$$
\begin{cases}
\mathbf{p} \in (H^1(B_R))^2, \\
b(\Omega_i; \mathbf{v}, \mathbf{p}) = \displaystyle\int_{\Gamma_{meas}} (\overline{\mathbf{u}_{meas} - \mathbf{u}}) \, \mathbf{v}, \qquad \forall \mathbf{v} \in (H^1(B_R))^2,
\end{cases}
\tag{74}
$$

the derivative $\dot{\mathbf{u}}$ is not needed and the shape derivative $DJ(\Omega) \cdot \mathbf{V}$ is given by the first term in (73). Problem (74) is the adjoint problem of (68).

The definition of Steklov–Poincaré operator implies that $L^*(\overline{\mathbf{p}}) = \overline{L(\mathbf{p})}$. Thus, the adjoint problem is equivalent to:

$$
\begin{cases}
\overline{\mathbf{p}} \in (H^1(B_R))^2, \\
b(\Omega_i; \overline{\mathbf{p}}, \mathbf{v}) = \displaystyle\int_{\Gamma_{meas}} (\overline{\mathbf{u}_{meas} - \mathbf{u}}) \, \overline{\mathbf{v}}, \qquad \forall \mathbf{v} \in (H^1(B_R))^2.
\end{cases}
\tag{75}
$$

Notice that $\mathbf{p}$ is then a solution of (66).

Step 4: Shape derivative. Let us calculate the first term in (73). For all $\mathbf{u}, \mathbf{p} \in (H^1(B_R))^2$

$$
\frac{d}{d\tau} b(\Omega_{i,\tau}; \mathbf{u}, \mathbf{p}) \Big|_{\tau=0} = \int_{\Omega'_e} \left(c_{j\alpha m\beta} \frac{\partial u_j}{\partial z_\alpha} \frac{\partial \overline{p}_m}{\partial z_\beta} - \rho\omega^2 \mathbf{u}\overline{\mathbf{p}} \right) \nabla \cdot \mathbf{V} \, d\mathbf{z}
$$

$$
- \int_{\Omega'_e} c_{j\alpha m\beta} \frac{\partial V_\gamma}{\partial z_\alpha} \frac{\partial u_j}{\partial z_\gamma} \frac{\partial \overline{p}_m}{\partial z_\beta} \, d\mathbf{z} - \int_{\Omega'_e} c_{j\alpha m\beta} \frac{\partial V_\gamma}{\partial z_\beta} \frac{\partial u_j}{\partial z_\alpha} \frac{\partial \overline{p}_m}{\partial z_\gamma} \, d\mathbf{z}.
$$

Integrating by parts, taking into account that $\mathbf{n}$ is the outward normal vector to Ω'_e and that $\mathbf{V}$ vanishes on Γ_R

$$
\frac{d}{d\tau} b(\Omega_{i,\tau}; \mathbf{u}, \mathbf{p}) \Big|_{\tau=0} = \int_\Gamma \left(c_{j\alpha m\beta} \frac{\partial u_j}{\partial z_\alpha} \frac{\partial \overline{p}_m}{\partial z_\beta} - \rho\omega^2 \mathbf{u}\overline{\mathbf{p}} \right) V_n \, dl
$$

$$
+ \int_{\Omega'_e} \left[c_{j\alpha m\beta} \left(\frac{\partial^2 u_j}{\partial z_\alpha \partial z_\gamma} \frac{\partial \overline{p}_m}{\partial z_\beta} + \frac{\partial u_j}{\partial z_\alpha} \frac{\partial^2 \overline{p}_m}{\partial z_\beta \partial z_\gamma} \right) - \rho\omega^2 \left(\frac{\partial \mathbf{u}}{\partial z_\gamma}\overline{\mathbf{p}} + \mathbf{u} \frac{\partial \overline{\mathbf{p}}}{\partial z_\gamma} \right) \right] V_\gamma \, d\mathbf{z}
$$

$$
- \int_{\Omega'_e} c_{j\alpha m\beta} \left(\frac{\partial^2 u_j}{\partial z_\alpha \partial z_\gamma} \frac{\partial \overline{p}_m}{\partial z_\beta} + \frac{\partial u_j}{\partial z_\gamma} \frac{\partial^2 \overline{p}_m}{\partial z_\beta \partial z_\alpha} + \frac{\partial^2 u_j}{\partial z_\alpha \partial z_\beta} \frac{\partial \overline{p}_m}{\partial z_\gamma} + \frac{\partial u_j}{\partial z_\alpha} \frac{\partial^2 \overline{p}_m}{\partial z_\beta \partial z_\gamma} \right) V_\gamma \, d\mathbf{z}
$$

$$
+ \int_\Gamma c_{j\alpha m\beta} \left(n_\alpha \frac{\partial u_j^-}{\partial z_\gamma} \frac{\partial \overline{p}_m^-}{\partial z_\beta} + n_\beta \frac{\partial u_j^-}{\partial z_\alpha} \frac{\partial \overline{p}_m^-}{\partial z_\gamma} \right) V_n n_\gamma \, dl.
\tag{76}
$$

The sum of all the integrals over Ω'_e that appear in (76) is

$$
- \int_{\Omega'_e} \left[\left(c_{j\alpha m\beta} \frac{\partial^2 u_j}{\partial z_\alpha \partial z_\beta} + \rho\omega^2 u_m \right) \frac{\partial \overline{p}_m}{\partial z_\gamma} + \left(c_{j\alpha m\beta} \frac{\partial^2 \overline{p}_m}{\partial z_\beta \partial z_\alpha} + \rho\omega^2 \overline{p}_j \right) \frac{\partial u_j}{\partial z_\gamma} \right] V_\gamma \, d\mathbf{z} = 0
$$

because

$$\rho\omega^2 u_j + c_{j\alpha m\beta}\frac{\partial^2 u_m}{\partial z_\alpha \partial z_\beta} = 0, \quad \rho\omega^2 \overline{p}_j + c_{j\alpha m\beta}\frac{\partial^2 \overline{p}_m}{\partial z_\alpha \partial z_\beta} = \overline{(u_{meas} - u)}_j\, \delta_{\Gamma_{meas}}, \quad \text{in } \Omega'_e$$

and $c_{j\alpha m\beta} = c_{m\beta j\alpha}$. The field $\mathbf{V} = 0$ on Γ_{meas}, thus all the integrals over Γ_{meas} are identically zero.

Adding the integrals on Γ we find

$$\int_\Gamma \left(c_{j\alpha m\beta}\frac{\partial u_j}{\partial z_\alpha}\frac{\partial \overline{p}_m}{\partial z_\beta} - \rho\omega^2 \mathbf{u}\overline{\mathbf{p}} \right) V_n\, dl$$

$$+ \int_\Gamma c_{j\alpha m\beta}\left(n_\alpha\frac{\partial u_j}{\partial z_\gamma}\frac{\partial \overline{p}_m}{\partial z_\beta} + n_\beta\frac{\partial u_j}{\partial z_\alpha}\frac{\partial \overline{p}_m}{\partial z_\gamma} \right) V_n n_\gamma\, dl.$$

Using the homogeneous Neumann condition, the second integral vanishes and (65) follows.   □

7.3 Topological Derivative

Once the shape derivative (65) is known, we use (6) to obtain

$$D_T(\mathbf{x}, \mathbb{R}^2) = \lim_{\varepsilon \to 0}\frac{-1}{\mathcal{V}'(\varepsilon)}\,\mathrm{Re}\left[\int_{\Gamma_\varepsilon}\left(c_{j\alpha m\beta}\frac{\partial u_j^\varepsilon}{\partial z_\alpha}\frac{\partial \overline{p}_m^\varepsilon}{\partial z_\beta} - \rho\omega^2 \mathbf{u}^\varepsilon \overline{\mathbf{p}}^\varepsilon \right)\, dl \right]. \quad (77)$$

Here, $\mathbf{u}^\varepsilon$ and $\mathbf{p}^\varepsilon$ are the solutions of the forward and adjoint Neumann problems (63) and (66) when $\Omega_e = \mathbb{R}^2 \setminus \overline{B}_\varepsilon(\mathbf{x})$ and $\Gamma = \Gamma_\varepsilon = \partial B_\varepsilon(\mathbf{x})$. As in the case of Helmholtz transmission problems, an asymptotic analysis of these solutions and their gradients at Γ_ε will help us to obtain the final expression for the topological derivative.

Theorem 7.2. *The topological derivative of the cost functional (64) is given by*

$$D_T(\mathbf{x}, \mathbb{R}^2) = \mathrm{Re}\left[a_{j\alpha m\beta}\frac{\partial u_j}{\partial z_\alpha}(\mathbf{x})\frac{\partial \overline{p}_m}{\partial z_\beta}(\mathbf{x}) - \rho\omega^2 \mathbf{u}(\mathbf{x})\overline{\mathbf{p}}(\mathbf{x}) \right] \quad (78)$$

where $\mathbf{u}$ and $\mathbf{p}$ solve the forward and adjoint Neumann problems with $\Omega_i = \emptyset$. The polarization tensor $a_{j\alpha m\beta}$ is defined by

$$a_{j\alpha m\beta} = c_{j\alpha m\beta} + c_{rsr's'}C_{pqj\alpha}C_{p'q'm\beta}\int_{|\boldsymbol{\xi}|=1}\frac{\partial \theta_r^{pq}}{\partial \xi_s}\frac{\partial \theta_{r'}^{p'q'}}{\partial \xi_{s'}}\,d\boldsymbol{\xi} \quad (79)$$

$$+ (c_{j\alpha rs}C_{pqm\beta} + c_{rsm\beta}C_{pqj\alpha})\int_{|\boldsymbol{\xi}|=1}\frac{\partial \theta_r^{pq}}{\partial \xi_s}\,d\boldsymbol{\xi}, \quad (80)$$

where $\mathbf{w}(\boldsymbol{\xi}) = c_{pqrs}\frac{\partial u_r}{\partial z_s}(\mathbf{x})\boldsymbol{\theta}^{pq}(\boldsymbol{\xi})$, and $\boldsymbol{\theta}^{pq}$ solve the problems

$$\begin{cases} c_{j\alpha m\beta}\dfrac{\partial^2 \theta_m^{pq}}{\partial \xi_\alpha \partial \xi_\beta} = 0, \quad j = 1,2, & \text{in } \mathbb{R}^2 \setminus \overline{B(0;1)}, \\[2mm] -c_{j\alpha m\beta}\dfrac{\partial \theta_m^{pq}}{\partial \xi_\beta}n_\alpha = M_{j\alpha}^{pq}n_\alpha, \quad j = 1,2, & \text{on } \Gamma, \end{cases} \quad (81)$$

*together with a radiation condition imposing decay at infinity. The matrices M^{pq}
are defined by*

$$M^{11} = \begin{pmatrix} 1 & 0 \\ 0 & 0 \end{pmatrix}, \quad M^{22} = \begin{pmatrix} 0 & 0 \\ 0 & 1 \end{pmatrix}, \quad M^{12} = \begin{pmatrix} 0 & 1/2 \\ 1/2 & 0 \end{pmatrix} = M^{21}.$$

In the 3D isotropic case, an explicit formula in terms of the elastic constants is
given in [25] using an explicit expression for $\mathbf{w}$. The 3D isotropic transmission
problem was studied in [26].

Proof. We expand $\mathbf{u}^\varepsilon(\mathbf{z}) = \mathbf{u}(\mathbf{z})\chi_{\mathbb{R}^2 \setminus \overline{B}_\varepsilon}(\mathbf{z}) + \mathbf{w}^\varepsilon(\mathbf{z})$, where $\mathbf{u} = \mathbf{u}_{inc}$ is the solution of the forward problem without obstacle ($\Omega_i = \emptyset$) and $\mathbf{w}^\varepsilon$ is the radiating
solution of

$$\begin{cases} \rho\omega^2 w_j^\varepsilon + c_{j\alpha m\beta}\dfrac{\partial^2 w_m^\varepsilon}{\partial z_\alpha \partial z_\beta} = 0, \quad j = 1, 2, & \text{in } \mathbb{R}^2 \setminus \overline{B}_\varepsilon, \\[3mm] -c_{j\alpha m\beta}\dfrac{\partial w_m^\varepsilon}{\partial z_\beta}n_\alpha = c_{j\alpha m\beta}\dfrac{\partial u_m}{\partial z_\beta}n_\alpha, \quad j = 1, 2, & \text{on } \Gamma_\varepsilon, \end{cases} \tag{82}$$

with Neumann data given by $\mathbf{u}$. The change of variable $\boldsymbol{\xi} = (\mathbf{z} - \mathbf{x})/\varepsilon$ transforms
this problem into:

$$\begin{cases} \varepsilon^2 \rho\omega^2 w_j^\varepsilon + c_{j\alpha m\beta}\dfrac{\partial^2 w_m^\varepsilon}{\partial \xi_\alpha \partial \xi_\beta} = 0, \quad j = 1, 2, & \text{in } \mathbb{R}^2 \setminus \overline{B}, \\[3mm] -c_{j\alpha m\beta}\dfrac{\partial w_m^\varepsilon}{\partial \xi_\beta}n_\alpha = \varepsilon c_{j\alpha m\beta}\dfrac{\partial u_m}{\partial z_\beta}n_\alpha, \quad j = 1, 2, & \text{on } \Gamma, \end{cases} \tag{83}$$

where B is the unit ball. As $\varepsilon \to 0$, we expand

$$\frac{\partial u_m(\mathbf{z})}{\partial z_\beta} = \frac{\partial u_m(\mathbf{x})}{\partial z_\beta} + O(\varepsilon)$$

and write $\mathbf{w}^\varepsilon = \mathbf{w}^{(1)} + \varepsilon \mathbf{w}^{(2)} + O(\varepsilon^2)$. The leading term, $\mathbf{w}^{(1)}$ is the radiating
solution of

$$\begin{cases} c_{j\alpha m\beta}\dfrac{\partial^2 w_m^{(1)}}{\partial \xi_\alpha \partial \xi_\beta} = 0, & \text{in } \mathbb{R}^2 \setminus \overline{B}, \\[3mm] -c_{j\alpha m\beta}\dfrac{\partial w_m^{(1)}}{\partial \xi_\beta}n_\alpha = 0, & \text{on } \Gamma. \end{cases} \tag{84}$$

The solution of this problem is $\mathbf{w}^{(1)}(\boldsymbol{\xi}) = 0$. The second term, $\mathbf{w}^{(2)}$, is the
radiating solution of

$$\begin{cases} c_{j\alpha m\beta}\dfrac{\partial^2 w_m^{(2)}}{\partial \xi_\alpha \partial \xi_\beta} = 0, & \text{in } \mathbb{R}^2 \setminus \overline{B}, \\[3mm] -c_{j\alpha m\beta}\dfrac{\partial w_m^{(2)}}{\partial \xi_\beta}n_\alpha = c_{j\alpha m\beta}\dfrac{\partial u_m}{\partial z_\beta}(\mathbf{x})n_\alpha, & \text{on } \Gamma. \end{cases} \tag{85}$$

Notice that, since the interface Γ is the unit circle, $\mathbf{n} = \boldsymbol{\xi}/|\boldsymbol{\xi}| = \boldsymbol{\xi}$. The function $\mathbf{w}^{(2)}$ can be written in terms of the elementary solutions of (81) as $\mathbf{w}^{(2)}(\boldsymbol{\xi}) = c_{pqrs}\frac{\partial u_r}{\partial z_s}(\mathbf{x})\boldsymbol{\theta}^{pq}(\boldsymbol{\xi})$. Our expansion yields:

$$\mathbf{u}^\varepsilon(\mathbf{z}) = \mathbf{u}(\mathbf{x}) + O(\varepsilon),$$

$$\frac{\partial u_m^\varepsilon}{\partial z_\beta}(\mathbf{z}) = \frac{\partial u_m}{\partial z_\beta}(\mathbf{x}) + c_{pqrs}\frac{\partial u_r}{\partial z_s}(\mathbf{x})\frac{\partial \theta_m^{pq}}{\partial \xi_\beta}(\boldsymbol{\xi}) + O(\varepsilon), \quad m,\beta = 1,2,$$

and similar expressions for $\overline{\mathbf{p}}$. Recalling that $\mathcal{V}'(\varepsilon) = -2\pi\varepsilon$, we find (78).   □

8 Conclusions

Topological derivative methods are a powerful tool to devise efficient numerical schemes for solving inverse scattering problems. In problems where the incident radiation on the obstacles is governed by Helmholtz or elasticity equations, we have computed expressions for the topological derivatives of appropriate constrained optimization regularizations. Topological derivative based numerical strategies to approximate the scatterers have been discussed. We have performed a number of tests in 2D and 3D illustrating the advantages and disadvantages of the method. In general, fairly good approximations of the number, size, location and shape of the scatterers are obtained at a low computational cost.

Many real situations require not only information about the scatterers, but also about constitutive parameters that we assume here to be known. An extension of topological derivative methods to provide information on both the obstacles and their parameters is proposed in [4].

Acknowledgements

This research has been supported by Grants MAT2005-05730-C02-02 and MTM-2004-01905 of the Spanish MEC, BSCH/UCM PR27/05-13939 and CM-910143.

References

1. Anagnostopoulos KA and Charalambopoulos A 2006 The linear sampling method for the transmission problem in two-dimensional anisotropic elasticity *Inverse problems* **22** 553–577
2. Bonnet M and Constantinescu A 2005 Inverse problems in elasticity *Inverse problems* **21** R1-R50
3. Burger M, Hackl B and Ring W 2004 Incorporating topological derivatives into level set methods *J. Comp. Phys.* **194** 344–362
4. Carpio A and Rapun ML, Solving inhomogeneous inverse problems by topological derivative methods, submitted, 2007
5. Carpio A and Rapun ML, Shape reconstruction in anisotropic elastic media. In preparation
6. Chandler GA and Sloan IH 1990. Spline qualocation methods for boundary integral equations. *Numer. Math.* **58**, 537–567

7. Charalambopoulos A 2002 On the interior transmission problem in nondissipative, homogeneous, anisotropic elasticity *J. Elast.* **67** 149–170

8. Colton D 1984 The inverse scattering problem for time-harmonic acoustic waves, SIAM Review **26** 323–350.

9. Colton D, Gieberman K and Monk P 2000 A regularized sampling method for solving three dimensional inverse scattering problems *SIAM J. Sci. Comput.* **21** 2316–2330

10. Colton D and Kress R 1983 *Integral equation methods in scattering theory* John Wiley & Sons. New York.

11. Colton D and Kress R 1992 *Inverse acoustic and electromagnetic scattering theory* Springer Berlin.

12. Colton D and Kirsch A 1996 A simple method for solving inverse scattering problems in the resonance region *Inverse problems* **12** 383–393

13. Costabel M and Stephan E 1985 A direct boundary integral equation method for transmission problems *J. Math. Anal. Appl.* **106** 367–413

14. Devaney AJ 1984 Geophysical diffraction tomography, *IEEE Trans. Geosci. Remote Sens.* **22** 3–13

15. Domínguez V, Rapún ML and Sayas FJ 2007 Dirac delta methods for Helmholtz transmission problems. To appear in *Adv. Comput. Math.*

16. Domínguez V and Sayas FJ 2003. An asymptotic series approach to qualocation methods. *J. Integral Equations Appl.* **15**, 113–151

17. Dorn O and Lesselier D 2006 Level set methods for inverse scattering *Inverse Problems* **22** R67–R131

18. Feijoo GR 2004 A new method in inverse scattering based on the topological derivative *Inverse Problems* **20** 1819–1840

19. Feijoo GR, Oberai AA and Pinsky PM 2004 An application of shape optimization in the solution of inverse acoustic scattering problems *Inverse problems* **20** 199–228

20. Gachechiladze A and Natroshvili D 2001 Boundary variational inequality approach in the anisotropic elasticity for the Signorini problem *Georgian Math. J.* 8 469-492

21. Garreau S, Guillaume P and Masmoudi M 2001 The topological asymptotic for PDE systems: the elasticity case *SIAM J. Control Optim.* **39** 1756–1778

22. Gegelia T and Jentsch L 1994 Potential methods in continuum mechanics *Georgian Math. J.* 599-640

23. Gerlach T and Kress R 1996 Uniqueness in inverse obstacle scattering with conductive boundary condition. *Inverse Problems* **12** 619–625

24. Guzina BB and Bonnet M 2006 Small-inclusion asymptotic of misfit functionals for inverse problems in acoustics *Inverse Problems* **22** 1761–1785

25. Guzina BB, Bonnet M 2004 Topological derivative for the inverse scattering of elastic waves *Q. Jl. Mech. Appl. Math.* **57** 161–179

26. Guzina BB, Chikichev I 2007 From imaging to material identification: A generalized concept of topological sensitivity *J. Mech. Phys. Sol.* **55** 245–279

27. Hettlich F 1995 Fréchet derivatives in inverse obstacle scattering *Inverse problems* **11** 371–382

28. Keller JB and Givoli D 1989 Exact non-reflecting boundary conditions *J. Comput. Phys.* **82** 172–192

29. Kirsch A, Kress R, Monk P and Zinn A 1988 Two methods for solving the inverse acoustic scattering problem *Inverse problems* **4** 749–770

30. Kirsch A and Kress R 1993 Uniqueness in inverse obstacle scattering *Inverse Problems* **9** 285–299

31. Kirsch A 1993 The domain derivative and two applications in inverse scattering theory *Inverse Problems* **9** 81–93

32. Kleinman RE and Martin P 1988 On single integral equations for the transmission problem of acoustics *SIAM J. Appl. Math* **48** 307–325

33. Kleinman RE and van der Berg PM 1992 A modified gradient method for two dimensional problems in tomography *J. Comput. Appl. Math.* **42** 17–35

34. Kress R and Roach GF 1978 Transmission problems for the Helmholtz equation *J. Math. Phys.* **19** 1433–1437

35. Kupradze VD, Gegelia TG, Basheleuishvili MO and Burchauladze TV, Three dimensional problems of the mathematical theory of elasticity and thermoelasticity, North-Holland Ser. Appl. Math. Mech. 25, North-Holland, Amsterdam, 1979.

36. Liseno A and Pierri R 2004 Imaging of voids by means of a physical optics based shape reconstruction algorithm *J. Opt. Soc. Am. A* **21** 968–974

37. Litman A, Lesselier D and Santosa F 1998 Reconstruction of a two-dimensional binary obstacle by controlled evolution of a level set *Inverse problems* **14** 685–706

38. Masmoudi M 1987 *Outils pour la conception optimale des formes* Thèse d'Etat en Sciences Mathématiques, Université de Nice

39. Masmoudi M, Pommier J and Samet B 2005 The topological asymptotic expansion for the Maxwell equations and some applications *Inverse Problems* **21** 547–564

40. Meddahi S and Sayas FJ 2005 Analysis of a new BEM–FEM coupling for two dimensional fluid–solid iteraction *Num. Methods Part. Diff. Eq.* **21** 1017–1154

41. Natterer F and Wubbeling F 1995 A propagation backpropagation method for ultrasound tomography *Inverse problems* **11** 1225-1232

42. Natroshvili D 1995 Mixed interface problems for anisotropic elastic bodies *Georgian Math. J.* 2 631-652

43. Natroshvili D 1996 Two-dimensional steady state oscillation problems of anisotropic elasticity *Georgian Math. J.* 3 239-262

44. Potthast R 1996 Fréchet differentiability of the solution to the acoustic Neumann scattering problem with respect to the domain *J. Inverse Ill-Posed Problems* **4** 67–84

45. Rapún ML and Sayas FJ 2006 Boundary integral approximation of a heat diffusion problem in time–harmonic regime *Numer. Algorithms* **41** 127–160

46. Rapún ML and Sayas FJ 2006 Indirect methods with Brakhage–Werner potentials for Helmholtz transmission problems. In Numerical Mathematics and advanced applications. ENUMATH 2005. Springer 1146–1154

47. Rapún ML and Sayas FJ 2007 Exterior Dirichlet and Neumann problems for the Hemholtz equation as limits of transmission problems. Submitted.

48. Rapún ML and Sayas FJ 2007 Boundary element simulation of thermal waves. Arch. Comput. Methods. Engrg **14** 3–46

49. Samet B, Amstutz S and Masmoudi M 2003 The topological asymptotic for the Helmholtz equation *SIAM J. Control Optimim* **42** 1523–1544

50. Santosa F 1996 A level set approach for inverse problems involving obstacles *ESAIM Control, Optim. Calculus Variations* **1** 17–33

51. Sloan IH 2000. Qualocation. *J. Comput. Appl. Math.* **125**, 461–478

52. Sokolowski J and Zolésio JP 1992 *Introduction to shape optimization. Shape sensitivity analysis* (Heidelberg: Springer)

53. Sokowloski J and Zochowski A 1999 On the topological derivative in shape optimization *SIAM J. Control Optim.* **37** 1251–1272

54. Torres RH and Welland GV 1993 The Helmholtz equation and transmission problems with Lipschitz interfaces *Indiana Univ. Math. J.* **42** 1457–1485

55. von Petersdorff T 1989 Boundary integral equations for mixed Dirichlet, Neumann and transmission problems. *Math. Methods Appl. Sci.* **11** 185–213

Time-Reversal and the Adjoint Imaging Method with an Application in Telecommunication

Oliver Dorn

Gregorio Millán Institute of Fluid Dynamics, Nanoscience and Industrial
Mathematics, Universidad Carlos III de Madrid, Leganés, Spain
odorn@math.uc3m.es

Summary. These lecture notes provide a mathematical treatment of time-reversal experiments with a special emphasis on telecommunication. A direct link is established between time-reversal experiments and the adjoint imaging method. Based on this relationship, several iterative schemes are proposed for optimizing MIMO (multiple-input multiple-output) time-reversal systems in underwater acoustic and in wireless communication systems. Whereas in typical imaging applications these iterative schemes require the repeated solution of forward problems in a computer, the analogue in time-reversal communication schemes consists of a small number of physical time-reversal experiments and does not require exact knowledge of the environment in which the communication system operates. The discussion is put in the general framework of wave propagation by symmetric hyperbolic systems, with detailed discussions of the linear acoustic system for underwater communication and of the time-dependent system of Maxwell's equations for telecommunication. Moreover, in its general form as treated here, the theory will also apply for several other models of wave-propagation, such as for example linear elastic waves.

1 Introduction

Time-reversal techniques have attracted great attention recently due to the large variety of interesting potential applications. The basic idea of time-reversal (often also referred to as 'phase-conjugation' in the frequency domain) can be roughly described as follows. A localized source emits a short pulse of acoustic, electromagnetic or elastic energy which propagates through a richly scattering environment. A receiver array records the arriving waves, typically as a long and complicated signal due to the complexity of the environment, and stores the time-series of measured signals in memory. After a short while, the receiver array sends the recorded signals time-reversed (i.e. first in – last out) back into the same medium. Due to the time-reversibility of the wave fields, the emitted energy backpropagates through the same environment, practically retracing the paths of the original signal, and refocuses, again as a short pulse, on the location where the source emitted the original pulse. This, certainly, is a slightly oversimplified description of the rather complex physics which is involved in the real time-reversal experiment. In practice, the quality of the refocused pulse depends on many parameters, as for example the randomness of the background medium, the size and location of the receiver array, temporal fluctuations of the environment, etc. Surprisingly, the quality of the refocused signal increases with increasing complexity of the background environment. This was observed experimentally by M. Fink and his group at the Laboratoire Ondes et Acoustique at Université Paris VII in Paris by a series of laboratory experiments (see for example [17,20,21]), and by W. A. Kuperman and co-workers at the Scripps Institution of Oceanography at University of California, San Diego, by a series of experiments performed between 1996 and 2000 in a shallow ocean environment (see for example [9,18,26]).

The list of possible applications of this time-reversal technique is long. In an iterated fashion, resembling the power method for finding maximal eigenvalues of a square matrix, the time-reversal method can be applied for focusing energy created by an ultrasound transducer array on strongly scattering objects in the region of interest. This can be used for example in lithotripsy for localizing and destroying gall-stones in an automatic way, or more generally in the application of medical imaging problems. Detection of cracks in the aeronautic industry, or of submarines in the ocean are other examples. See for example [39,40], and for related work [7,12]. The general idea of time-reversibility, and its use in imaging and detection, is certainly not that new. Looking into the literature for example of seismic imaging, the application of this basic idea can be found in a classical and very successful imaging strategy for detecting scattering interfaces in the Earth, the so-called 'migration' technique [4,8]. However, the systematic use and investigation of the time-reversal phenomenon and its experimental realizations started more recently, and has been carried out during the last 10–15 years or so by different research groups. See for example [17,18,20,21,26,31,46,49] for experimental demonstrations, and [1–3,5,6,9,15,16,23,28,30,38,39,43–45,47,50] for theoretical and numerical approaches.

One very young and promising application of time-reversal is *communication*. In these lecture notes we will mainly concentrate on that application, although the general results should carry over also to other applications as mentioned above.

The text is organized as follows. In Sect. 2 we give a very short introduction into time-reversal in the ocean, with a special emphasis on underwater sound communication. Wireless communication in a MIMO setup, our second main application in these notes, is briefly presented in Sect. 3. In Sect. 4 we discuss symmetric hyperbolic systems in the form needed here, and examples of such systems are given in Sect. 5. In Sect. 6, the basic spaces and operators necessary for our mathematical treatment are introduced. The inverse problem in communication, which we are focusing on in these notes, is then defined in Sect. 7. In Sect. 8, we derive the basic iterative scheme for solving this inverse problem. Section 9 gives practical expressions for calculating the adjoint communication operator, which plays a key role in the iterative time-reversal schemes. In Sect. 10 the acoustic time-reversal mirror is defined, which will provide the link between the 'acoustic time-reversal experiment' and the adjoint communication operator. The analogous results for the electromagnetic time-reversal mirror are discussed in Sect. 11. Section 12 combines the results of these two sections, and explicitly provides the link between time-reversal and the adjoint imaging method. Sections 13, 14, and 15 propose then several different iterative time-reversal schemes for solving the inverse problem of communication, using this key relationship between time-reversal and the adjoint communication operator. The practically important issue of partial measurements (and generalized measurements) is treated in Sect. 16. Finally, Sect. 17 summarizes the results of these lecture notes, and points out some interesting future research directions.

2 Time-Reversal and Communication in the Ocean

The ocean is a complex wave-guide for sound [10]. In addition to scattering effects at the top and the bottom of the ocean, also its temperature profile and the corresponding refractive effects contribute to this wave-guiding property and allow acoustic energy to travel large distances. Typically, this propagation has a very complicated behaviour. For example, multipathing occurs if source and receiver of sound waves are far away from each other, since due to scattering and refraction there are many possible paths on which acoustic energy can travel between them. Surface waves and air bubbles at the top of the ocean, sound propagation through the rocks and sedimentary layers at the bottom of the ocean, and other effects further contribute to the complexity of sound propagation in the ocean. When a source (e.g. the base station of a communication system in the ocean) emits a short signal at a given location, the receiver (e.g. a user of this communication system) some distance away from the source typically receives a long and complicated signal due to the various influences of the ocean to this signal along the different connecting paths. If the base station wants to communicate with the user by sending a series of short signals, this complex response of the ocean to the signal needs to be resolved and taken into account.

In a classical communication system, the base station which wants to communicate with a user broadcasts a series of short signals (e.g. a series of 'zeros' and 'ones') into the environment. The hope is that the user will receive this message as a similar series of temporally well-resolved short signals which can be easily identified and decoded. However, this almost never occurs in a complex environment, due to the multipathing and the resulting delay-spread of the emitted signals. Typically, when the base station broadcasts a series of short signals into such an environment, intersymbol interference occurs at the user position due to the temporal overlap of the multipath contributions of these signals. In order to recover the individual signals, a significant amount of signal processing is necessary at the user side, and, most importantly, the user needs to have some knowledge of the propagation behaviour of the signals in this environment (i.e. he needs to know the 'channel'). Intersymbol interference can in principle be avoided by adding a sufficient delay between individual signals emitted at the base station which takes into account the delay-spread in the medium. That, however, slows down the communication, and reduces the capacity of the environment as a communication system. An additional drawback of simply broadcasting communication signals from the base station is the obvious lack of interception security. A different user of the system who also knows the propagation behaviour of signals in the environment, can equally well resolve the series of signals and decode them.

Several approaches have been suggested to circumvent the above mentioned drawbacks of communication in multiple-scattering environments. Some very promising techniques are based on the time-reversibility property of propagating wave-fields [18, 19, 29, 32, 34, 35, 37, 44]. The basic idea is as follows. The user who wants to communicate with the base station, starts the communication process by sending a short pilot signal through the environment. The base station receives this signal as a long and complex signal due to multipathing. It time-reverses the received signal and sends it back into the environment. The backpropagating waves will produce a complicated wave-field everywhere due to the many interfering parts of the emitted signal. However, due to the time-reversibility of the wave-fields, one expects that the interference will be constructive at the position of the user who sent the original pilot signal, and mainly destructive at all other positions. Therefore, the user will receive a short signal very similar to (ideally, a time-reversed replica of) the originally sent pilot signal. All other users who might be in the environment at the same time will only receive noise speckle due to incoherently interfering contributions of the backpropagating field. If the base station sends the individual elements ('ones' and 'zeros') of the intended message in a phase-encoded form as a long overlapping string of signals, the superposition principle will ensure that, at the user position, this string of signals will appear as a series of short well-separated signals, each resembling some phase-shifted (and time-reversed) form of the pilot signal.

In order to find out whether this theoretically predicted scenario actually takes place in a real multiple-scattering environment like the ocean, Kuperman et al. have performed a series of four experiments in a shallow ocean environment between 1996 and 2000, essentially following the above described scenario. The experiments have been performed at a Mediterranean location close to the Italian coast. (A similar setup was also used in an experiment performed in 2001 off the coast of New England which has been reported in Yang [49].) A schematic view of these experiments is shown in Fig. 1. The single 'user' is replaced here by a 'probe source', and the 'source-receive array' (SRA) plays the role of the 'base station'. An additional vertical receive array (VRA) was deployed at the position of the probe source in order to measure the temporal and spatial spread of the backpropagating fields in the neighbourhood of the probe source location. In this shallow environment (depths of about 100–200 m, and distances between 10 and 30 km) the multipathing of the waves is mostly caused by multiple reflections at the surface and the bottom of the ocean. The results of the experiments have been reported in [18, 26, 31]. They show that in fact a strong spatial and temporal focusing of the backpropagating waves occurs at the source position.

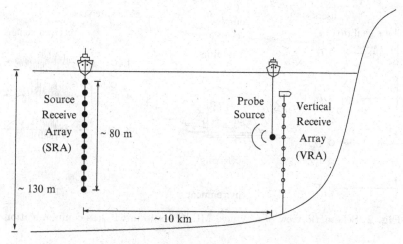

Fig. 1. The communication problem in underwater acoustics

A theoretical explanation of the temporal and spatial refocusing of time-reversed waves in random environments has been given in Blomgren et al. [5].

3 The MIMO Setup in Wireless Communication

The underwater acoustics scenario described above directly carries over to situations which might be more familiar to most of us, namely to the more and more popular wireless communication networks using mainly electromagnetic waves in the microwave regime. Starting from the everyday use of cell-phones, ranging to small wireless-operating local area networks (LAN) for computer systems or for private enterprise communication systems, exactly the same problems arise as in underwater communication. The typically employed microwaves of a wavelength at about 10–30 cm are heavily scattered by environmental objects like cars, fences, trees, doors, furniture, etc. This causes a very complicated multipath structure of the signals received by users of such a communication system. Since bandwidths are limited and increasingly expensive, a need for more and more efficient communication systems is imminent. Recently, the idea of a so-called multiple-input multiple-output (MIMO) communication system has been introduced with the potential to increase the capacity and efficiency of wireless communication systems [22]. The idea is to replace a single antenna at the base station which is responsible for multiple users, or even a system where one user communicates with just one dedicated base antenna, by a more general system where an array of multiple antennas at the base station is interacting simultaneously and in a complex way with multiple users. A schematic description of such a MIMO system (with seven base antennas and seven users) is given in Fig. 2. See for example [24, 42] for recent overviews on MIMO technology.

Time-reversal techniques are likely to play also here a key role in improving communication procedures and for optimizing the use of the limited resources

140   O. Dorn

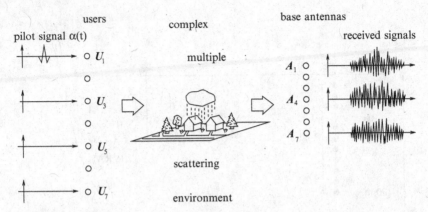

Fig. 2. Schematic view of a typical MIMO setup in wireless communication

(especially bandwidth) which are available for this technology [19, 29, 32, 37]. One big advantage of time-reversal techniques is that they are automatically adapted to the complex environment and that they can be very fast since they do not require heavy signal processing at the receiver or user side. In [34, 35], an *iterated time-reversal scheme* for the optimal refocusing of signals in such a MIMO communication system was proposed, which we will describe in more details in Sect. 13.

We will establish a direct link between the time-reversal technique and solution strategies for inverse problems. As an application of this relationship, we will derive iterative time-reversal schemes for the optimization of wireless or underwater acoustic MIMO communication systems. The derivation is performed completely in time-domain, for very general first order symmetric hyperbolic systems describing wave propagation phenomena in a complex environment. One of the schemes which we derive, in a certain sense the 'basic one', will turn out to be practically equivalent to the scheme introduced in [34, 35], although the derivation uses different tools. Therefore, we provide a new interpretation of that scheme. The other schemes which we introduce are new in this application, and can be considered as either generalizations of the basic scheme, or as independent alternatives to that scheme. Each of them addresses slightly different objectives and has its own very specific characteristics.

4 Symmetric Hyperbolic Systems

We treat wave propagation in communication systems in the general framework of symmetric hyperbolic systems of the form

$$\Gamma(\mathbf{x})\frac{\partial \mathbf{u}}{\partial t} + \sum_{i=1}^{3} D^i \frac{\partial \mathbf{u}}{\partial x_i} + \Phi(\mathbf{x})\mathbf{u} = \mathbf{q} \tag{1}$$

$$\mathbf{u}(\mathbf{x}, 0) = 0. \tag{2}$$

Here, $\mathbf{u}(\mathbf{x}, t)$ and $\mathbf{q}(\mathbf{x}, t)$ are real-valued time-dependent N-vectors, $\mathbf{x} \in \mathbb{R}^3$, and $t \in [0, T]$. $\Gamma(\mathbf{x})$ is a real, symmetric, uniformly positive definite $N \times N$-matrix, i.e. $\Gamma(\mathbf{x}) \geq \epsilon$ for some $\epsilon > 0$. Moreover, $\Phi(\mathbf{x})$ is a symmetric positive semi-definite $N \times N$ matrix, i.e. $\Phi(\mathbf{x}) \geq 0$. It models possible energy loss through dissipation in the medium. The D^i are real, symmetric and independent of $(\mathbf{x}, t)$. We will also use the short notation

$$\Lambda := \sum_{i=1}^{3} D^i \frac{\partial}{\partial x^i}. \tag{3}$$

In addition to the above mentioned assumptions on the coefficients $\Gamma(\mathbf{x})$ and $\Phi(\mathbf{x})$, we will assume throughout this text that all quantities $\Gamma(\mathbf{x})$, $\Phi(\mathbf{x})$, $\mathbf{u}(\mathbf{x}, t)$ and $\mathbf{q}(\mathbf{x}, t)$ are 'sufficiently regular' in order to safely apply for example integration by parts and Green's formulas. For details see for example [11, 33]. We will assume that no energy reaches the boundaries $\partial\Omega$ during the time $[0, T]$, such that we will always have

$$\mathbf{u}(\mathbf{x}, t) = 0 \quad \text{on} \quad \partial\Omega \times [0, T]. \tag{4}$$

The *energy density* $\mathcal{E}(\mathbf{x}, t)$ is defined by

$$\mathcal{E}(\mathbf{x}, t) = \frac{1}{2} \langle \Gamma(\mathbf{x})\mathbf{u}(\mathbf{x}, t), \mathbf{u}(\mathbf{x}, t) \rangle_N$$

$$= \frac{1}{2} \sum_{m,n=1}^{N} \Gamma_{mn}(\mathbf{x})\mathbf{u}_m(\mathbf{x}, t)\mathbf{u}_n(\mathbf{x}, t).$$

The *total energy* $\hat{\mathcal{E}}(t)$ in Ω at a given time t is therefore

$$\hat{\mathcal{E}}(t) = \frac{1}{2} \int_{\Omega} \langle \Gamma(\mathbf{x})\mathbf{u}(\mathbf{x}, t), \mathbf{u}(\mathbf{x}, t) \rangle_N \, d\mathbf{x}.$$

The *flux* $\mathcal{F}(\mathbf{x}, t)$ is given by

$$\mathcal{F}_i(\mathbf{x}, t) = \frac{1}{2} \langle D^i \mathbf{u}(\mathbf{x}, t), \mathbf{u}(\mathbf{x}, t) \rangle_N \quad , \quad i = 1, 2, 3.$$

5 Examples for Symmetric Hyperbolic Systems

In the following, we want to give some examples for symmetric hyperbolic systems as defined above. Of special interest for communication are the system of the linearized acoustic equations and the system of Maxwell's equations. We will discuss these two examples in detail in these lecture notes. Another important example for symmetric hyperbolic systems is the system of elastic waves equations, which we will however leave out in our discussion for the sake of brevity. We only mention that all the results derived here apply without restrictions also to linear elastic waves. Elastic wave propagation becomes important for example in ocean acoustic communication models which incorporate wave propagation through the sedimentary and rock layers at the bottom of the ocean.

5.1 Linearized Acoustic Equations

As a model for *underwater sound propagation*, we consider the following linearized form of the acoustic equations in an isotropic medium

$$\rho(\mathbf{x})\frac{\partial \mathbf{v}}{\partial t} + \operatorname{grad} p = \mathbf{q}_v \tag{5}$$

$$\kappa(\mathbf{x})\frac{\partial p}{\partial t} + \operatorname{div} \mathbf{v} = \mathbf{q}_p \tag{6}$$

$$p(\mathbf{x},0) = 0, \qquad \mathbf{v}(\mathbf{x},0) = 0. \tag{7}$$

Here, $\mathbf{v}$ is the velocity, p the pressure, ρ the density, and κ the compressibility. We have $N = 4$, $\mathbf{u} = (\mathbf{v}, p)^T$ and $\mathbf{q} = (\mathbf{q}_v, \mathbf{q}_p)^T$ (where 'T' as a superscript always means 'transpose'). Moreover, we have

$$\Gamma(\mathbf{x}) = \operatorname{diag}(\rho(x), \rho(x), \rho(x), \kappa(x)) \quad \text{and} \quad \Phi(\mathbf{x}) = 0.$$

With the notation $\varphi = (\partial_1, \partial_2, \partial_3)^T$, we can write Λ as

$$\Lambda = \begin{pmatrix} 0 & \varphi \\ \varphi^T & 0 \end{pmatrix}.$$

The operators D^i, $i = 1, 2, 3$, can be recovered from Λ by putting

$$D^i_{m,n} = \begin{cases} 1 \text{ where } \Lambda_{m,n} = \partial_i, \\ 0 \text{ elsewhere} \end{cases}$$

The energy density $\mathcal{E}(\mathbf{x}, t)$ is given by

$$\mathcal{E}(\mathbf{x},t) = \frac{1}{2}\big(\rho(\mathbf{x})|\mathbf{v}(\mathbf{x},t)|^2 + \kappa(\mathbf{x})p^2(\mathbf{x},t)\big),$$

and the energy flux $\mathcal{F}(\mathbf{x}, t)$ is

$$\mathcal{F}(\mathbf{x},t) = p(\mathbf{x},t)\mathbf{v}(\mathbf{x},t).$$

We mention that the dissipative case $\Phi(\mathbf{x}) \neq 0$ can be treated as well in our framework, and yields analogous results to those presented here.

5.2 Maxwell's Equations

As a second example, we will consider Maxwell's equations for an anisotropic medium with some energy loss due to the inherent conductivity. This can for example model *wireless communication* in a complex environment.

$$\epsilon(\mathbf{x})\frac{\partial \mathbf{E}}{\partial t} - \operatorname{curl} \mathbf{H} + \sigma(\mathbf{x})\mathbf{E} = \mathbf{q}_E \tag{8}$$

$$\mu(\mathbf{x})\frac{\partial \mathbf{H}}{\partial t} + \operatorname{curl} \mathbf{E} = \mathbf{q}_H \tag{9}$$

$$\mathbf{E}(\mathbf{x}, 0) = 0, \qquad \mathbf{H}(\mathbf{x}, 0) = 0. \tag{10}$$

We have $N = 6$, and $\mathbf{u} = (\mathbf{E}, \mathbf{H})^T$, $\mathbf{q} = (\mathbf{q}_E, \mathbf{q}_H)^T$. Moreover,

$$\Gamma(\mathbf{x}) = \mathbf{diag}(\epsilon(\mathbf{x}), \mu(\mathbf{x}))$$
$$\Phi(\mathbf{x}) = \mathbf{diag}(\sigma(\mathbf{x}), 0).$$

Here, ϵ and μ are symmetric positive definite 3×3 matrices modelling the anisotropic permittivity and permeability distribution in the medium, and σ is a symmetric positive semi-definite 3×3 matrix which models the anisotropic conductivity distribution. In wireless communication, this form can model for example dissipation by conductive trees, conductive wires, rainfall, pipes, etc. The operator Λ can be written in block form as

$$\Lambda = \begin{pmatrix} 0 & -\Xi \\ \Xi & 0 \end{pmatrix},$$

with

$$\Xi = \begin{pmatrix} 0 & -\partial_3 & \partial_2 \\ \partial_3 & 0 & -\partial_1 \\ -\partial_2 & \partial_1 & 0 \end{pmatrix}.$$

The operators D^i, $i = 1, 2, 3$, can be recovered from Λ by putting

$$D^i_{m,n} = \begin{cases} 1 & \text{where} \quad \Lambda_{m,n} = \partial_i, \\ -1 & \text{where} \quad \Lambda_{m,n} = -\partial_i, \\ 0 & \text{elsewhere} \end{cases}$$

The energy density $\mathcal{E}(\mathbf{x}, t)$ is given by

$$\mathcal{E}(\mathbf{x}, t) = \frac{1}{2}\left(\epsilon(\mathbf{x})|\mathbf{E}(\mathbf{x}, t)|^2 + \mu(\mathbf{x})|\mathbf{H}(\mathbf{x}, t)|^2\right).$$

The energy flux $\mathcal{F}(\mathbf{x}, t)$ is described by the *Poynting vector*

$$\mathcal{F}(\mathbf{x}, t) = \mathbf{E}(\mathbf{x}, t) \times \mathbf{H}(\mathbf{x}, t).$$

5.3 Elastic Waves Equations

As already mentioned, also elastic waves can be treated in the general framework of symmetric hyperbolic systems. For more details we refer to [33, 41].

6 The Basic Spaces and Operators

For our mathematical treatment of time-reversal we introduce the following spaces and inner products. We assume that we have J users $\mathcal{U}_j$, $j = 1, \ldots, J$ in our system, and in addition a base station which consists of K antennas $\mathcal{A}_k$, $k = 1, \ldots, K$. Each user and each antenna at the base station can receive, process and emit signals which we denote by $\mathbf{s}_j(t)$ for a given user $\mathcal{U}_j$, $j = 1, \ldots, J$, and by $\mathbf{r}_k(t)$ for a given base antenna $\mathcal{A}_k$, $k = 1, \ldots, K$. Each of these signals consists of a time-dependent N-vector indicating measured or processed signals of time-length T. In our analysis we will often have to consider functions defined on the time interval $[T, 2T]$ instead of $[0, T]$. For simplicity, we will use the same notation for the function spaces defined on $[T, 2T]$ as we use for those defined on $[0, T]$. It will always be obvious which space we refer to in a given situation.

Lumping together all signals at the users on the one hand, and all signals at the base station on the other hand, yields the two fundamental quantities

$$\mathbf{s} = (\mathbf{s}_1, \ldots, \mathbf{s}_J) \in \hat{Z}$$
$$\mathbf{r} = (\mathbf{r}_1, \ldots, \mathbf{r}_K) \in Z$$

with

$$\hat{Z} = \left(L_2([0, T])^N\right)^J, \qquad Z = \left(L_2([0, T])^N\right)^K.$$

The two signal spaces Z and $\hat{Z}$ introduced above are equipped with the inner products

$$\left\langle \mathbf{s}^{(1)}, \mathbf{s}^{(2)} \right\rangle_{\hat{Z}} = \sum_{j=1}^{J} \int_{[0,T]} \left\langle \mathbf{s}_j^{(1)}(t), \mathbf{s}_j^{(2)}(t) \right\rangle_N dt$$

$$\left\langle \mathbf{r}^{(1)}, \mathbf{r}^{(2)} \right\rangle_{Z} = \sum_{k=1}^{K} \int_{[0,T]} \left\langle \mathbf{r}_k^{(1)}(t), \mathbf{r}_k^{(2)}(t) \right\rangle_N dt.$$

The corresponding norms are

$$\|\mathbf{s}\|_{\hat{Z}}^2 = \langle \mathbf{s}, \mathbf{s} \rangle_{\hat{Z}}, \qquad \|\mathbf{r}\|_Z^2 = \langle \mathbf{r}, \mathbf{r} \rangle_Z$$

Each user $\mathcal{U}_j$ and each antenna $\mathcal{A}_k$ at the base station can send a given signal $\mathbf{s}_j(t)$ or $\mathbf{r}_k(t)$, respectively. This gives rise to a source distribution $\hat{\mathbf{q}}_j(\mathbf{x}, t)$, $j = 1, \ldots, J$, or $\mathbf{q}_k(\mathbf{x}, t)$, $k = 1, \ldots, K$, respectively. Here and in the following we will use in our notation the following convention. If one symbol appears in both forms, with and without a 'hat' ($\hat{}$) on top of this symbol, then all quantities with the 'hat' symbol are related to the users, and those without the 'hat' symbol to the antennas at the base station.

Each of the sources created by a user or by a base antenna will appear on the right hand side of (1) as a mathematical source function and gives rise to a corresponding wave field which satisfies (1), (2). When solving the system (1), (2), typically certain Sobolev spaces need to be employed for the appropriate description of the underlying function spaces (see for example [11]). For our purposes, however, it will be sufficient to assume that both, source functions and wave fields, are members of the following canonical function space U which is defined as

$$U = \{\mathbf{u} \in L^2(\Omega \times [0,T])^N, \ \mathbf{u} = 0 \text{ on } \partial\Omega \times [0,T], \ \|\mathbf{u}\|_U < \infty\},$$

and which we have equipped with the usual energy inner product

$$\langle \mathbf{u}(t), \mathbf{v}(t) \rangle_U = \int_{[0,T]} \int_\Omega \langle \Gamma(\mathbf{x})\mathbf{u}(\mathbf{x},t), \mathbf{v}(\mathbf{x},t) \rangle_N \, d\mathbf{x}dt,$$

and the corresponding energy norm

$$\|\mathbf{u}\|_U^2 = \langle \mathbf{u}, \mathbf{u} \rangle_U.$$

Also here, in order to simplify the notation, we will use the same space when considering functions in the shifted time interval $[T, 2T]$ instead of $[0, T]$.

Typically, when a user or an antenna at the base station transforms a signal into a source distribution, it is done according to a very specific antenna characteristic which takes into account the spatial extension of the user or the antenna. We will model this characteristic at the user by the functions $\hat{\gamma}_j(\mathbf{x})$, $j = 1, \ldots, J$, and for base antennas by the functions $\gamma_k(\mathbf{x})$, $k = 1, \ldots, K$. With these functions, we can introduce the linear 'source operators' $\hat{Q}$ and Q mapping signals $\mathbf{s}$ at the set of users and $\mathbf{r}$ at the set of base antennas into the corresponding source distributions $\hat{\mathbf{q}}(\mathbf{x},t)$ and $\mathbf{q}(\mathbf{x},t)$, respectively. They are given as

$$\hat{Q} : \hat{Z} \mapsto U, \quad \hat{\mathbf{q}}(\mathbf{x},t) = \hat{Q}\mathbf{s} = \sum_{j=1}^{J} \hat{\gamma}_j(\mathbf{x})\mathbf{s}_j(t),$$

$$Q : Z \to U, \quad \mathbf{q}(\mathbf{x},t) = Q\mathbf{r} = \sum_{k=1}^{K} \gamma_k(\mathbf{x})\mathbf{r}_k(t).$$

We will assume that the functions $\hat{\gamma}_j(\mathbf{x})$ are supported on a small neighbourhood $\hat{V}_j$ of the user location $\hat{\mathbf{d}}_j$, and that the functions $\gamma_k(\mathbf{x})$ are supported on a small neighbourhood V_k of the antenna location $\mathbf{d}_k$. Moreover, all these neighbourhoods are strictly disjoint to each other. For example, the functions $\hat{\gamma}_j(\mathbf{x})$ could be assumed to be L_2-approximations of the Dirac delta measure $\delta(\mathbf{x} - \hat{\mathbf{d}}_j)$ concentrated at the user locations $\hat{\mathbf{d}}_j$, and the functions $\gamma_k(\mathbf{x})$ could be assumed to be L_2-approximations of the Dirac delta measure $\delta(\mathbf{x} - \mathbf{d}_k)$ concentrated at the antenna locations $\mathbf{d}_k$.

Both, users and base antennas can also record incoming fields $\mathbf{u} \in U$ and transform the recorded information into signals. Also here, this is usually done according to very specific antenna characteristics of each user and each base antenna. For simplicity (and without loss of generality), we will assume that the antenna characteristic of a user or base antenna for receiving signals is the same as for transmitting signals, namely $\hat{\gamma}_j(\mathbf{x})$ for the user and $\gamma_k(\mathbf{x})$ for a base antenna. (The case of more general source and measurement operators is discussed in Sect. 16.) With this, we can define the linear 'measurement operators' $\hat{M} : U \to \hat{Z}$ and $M : U \to Z$, respectively, which transform incoming fields into measured signals, by

$$\mathbf{s}_j(t) = (\hat{M}\mathbf{u})_j = \int_\Omega \hat{\gamma}_j(\mathbf{x})\mathbf{u}(\mathbf{x},t)d\mathbf{x}, \quad (j = 1,\dots,J)$$

$$\mathbf{r}_k(t) = (M\mathbf{u})_k = \int_\Omega \gamma_k(\mathbf{x})\mathbf{u}(\mathbf{x},t)d\mathbf{x}, \quad (k = 1,\dots,K)$$

Finally, we define the linear operator F mapping sources $\mathbf{q}$ to states $\mathbf{u}$ by

$$F : U \to U, \qquad F\mathbf{q} = \mathbf{u},$$

where $\mathbf{u}$ solves the problem (1), (2). As already mentioned, we assume that the domain Ω is chosen sufficiently large and that the boundary $\partial\Omega$ is sufficiently far away from the users and base antennas, such that there is no energy reaching the boundary in the time interval $[0, T]$ (or $[T, 2T]$) due to the finite speed of signal propagation. Therefore, the operator F is well-defined.

Formally, we can now introduce the two linear *communication operators* A and B. They are defined as

$$A : Z \to \hat{Z}, \qquad A\mathbf{r} = \hat{M}FQ\mathbf{r},$$
$$B : \hat{Z} \to Z, \qquad B\mathbf{s} = MF\hat{Q}\mathbf{s}.$$

The operator A models the following situation. The base station emits the signal $\mathbf{r}(t)$ which propagates through the complex environment. The users measure the arriving wave fields and transform them into measurement signals. The measured signals at the set of all users is $\mathbf{s}(t) = A\mathbf{r}(t)$. The operator B describes exactly the reversed situation. All users emit together the set of signals $\mathbf{s}(t)$, which propagate through the given complex environment and are received by the base station. The corresponding set of measured signals at all antennas of the base station is just $\mathbf{r}(t) = B\mathbf{s}(t)$. No time-reversal is involved so far.

7 An Inverse Problem Arising in Communication

In the following, we outline a typical problem arising in communication, which gives rise to a mathematically well-defined inverse problem.

A specified user of the system, say $\mathcal{U}_1$, defines a (typically but not necessarily short) pilot signal $\alpha(t)$ with the goal to use it as a template for receiving the information from the base station. The base station wants to emit a signal $\tilde{\mathbf{r}}(t)$ which, after having travelled through the complex environment and arriving at the user $\mathcal{U}_1$, matches this pilot signal as closely as possible. Neither the base station nor any other user except of $\mathcal{U}_1$ are required (or expected) to know the correct form of the pilot signal $\alpha(t)$ for this problem. As an additional constraint, the base station wants that at the other users $\mathcal{U}_j$, $j > 1$, as little energy as possible arrives when communicating with the specified user $\mathcal{U}_1$. This is also in the interest of the other users, who want to use a different 'channel' for communicating at the same time with the base antenna, and want to minimize interference with the communication initiated by user $\mathcal{U}_1$. The complex environment itself in which the communication takes place (i.e. the 'channel') is assumed to be unknown to all users and to the base station.

In order to arrive at a mathematical description of this problem, we define the 'ideal signal' $\tilde{s}(t)$ received by all users as

$$\tilde{s}(t) = (\alpha(t), 0, \ldots, 0)^T. \tag{11}$$

Each user only knows its own component of this signal, and the base antenna does not need to know any component of this ideal signal at all.

Definition 7.1. The inverse problem of communication: In the terminology of inverse problems, the above described scenario defines an inverse source problem, which we call for the purpose of these lecture notes the 'inverse problem of communication'. The goal is to find a 'source distribution' $\tilde{r}(t)$ at the base station which satisfies the 'data' $\tilde{s}(t)$ at the users:

$$A\tilde{r} = \tilde{s}. \tag{12}$$

The 'state equation' relating sources to data is given by the symmetric hyperbolic system (1), (2).

Remark 7.1. Notice that the basic operator A in (12) is unknown to the users and the base station since they typically do not know the complicated medium in which the waves propagate. If the operator A (together with $\tilde{s}$) would be known at the base station by some means, the inverse source problem formulated above could be solved using classical inverse problems techniques, which would be computationally expensive but in principle doable. In the given situation, the user and the base station are able to do physical experiments, which amounts to 'applying' the communication operator A to a given signal. Determining the operator A explicitly by applying it to a set of basis functions of Z would be possible, but again it would be too expensive. We will show in the following that, nevertheless, many of the classical solution schemes known from inverse problems theory can be applied in this situation even without knowing the operator A explicitly. The basic tool which we will use is a series of time-reversal experiments, applied to carefully designed signals at the users and the base station.

Remark 7.2. A practical template for an *iterative scheme for finding an optimal signal* at the base station can be envisioned as follows. User $\mathcal{U}_1$ starts the communication process by emitting an initial signal $s_1^{(0)}(t)$ into the complex environment. This signal, after having propagated through the complex environment, finally arrives at the base station and is received there usually as a relatively long and complicated signal due to the multiple scattering events it experienced on its way. When the base station receives such a signal, it processes it and sends a new signal $\mathbf{r}^{(1)}(t)$ back through the environment which is received by all users. After receiving this signal, all users become active. The user $\mathcal{U}_1$ compares the received signal with the pilot signal. If the match is not good enough, the received signal is processed by this user in order to optimize the match with the pilot signal when receiving the next iterate from the base station. All other users identify the received signal as unwanted noise, and process it with the goal to receive in the next iterate from the base station a signal with lower amplitude, such that it does not interfere with their own communications. All users send now their processed signals, which together define $\mathbf{s}^{(1)}(t)$, back to the base station. The base station receives them all simultaneously, again usually as a long and complicated signal, processes this signal and sends a new signal $\mathbf{r}^{(2)}(t)$ back into the environment which ideally will match the desired signals at all users better than the previously emitted signal $\mathbf{r}^{(1)}(t)$. This iteration stops when all users are satisfied, i.e. when user $\mathcal{U}_1$ receives a signal which is sufficiently close to the pilot signal $\alpha(t)$, and the energy or amplitude of the signals arriving at the other users has decreased enough in order not to disturb their own communications. After this learning process of the channel has been completed, the user $\mathcal{U}_1$ can now start communicating safely with the base antenna using the chosen pilot signal $\alpha(t)$ for decoding the received signals.

Similar schemes have been suggested in [18, 19, 29, 32, 37, 44] in a single-step fashion, and in [34, 35] performing multiple steps of the iteration. The main questions to be answered are certainly which signals each user and each base antenna needs to emit in each step, how these signals need to be processed, at which stage this iteration should be terminated, and which optimal solution this scheme is expected to converge to. *One of the main objectives of these lecture notes is to provide a theoretical framework for answering these questions by combining basic concepts of inverse problems theory with experimental time-reversal techniques.*

8 The Basic Approach for Solving the Inverse Problem of Communication

A standard approach for solving problem (12) in the situation of noisy data is to look for the least-squares solution

$$\mathbf{r}_{LS} = \text{Min}_{\mathbf{r}} \| A\mathbf{r} - \tilde{\mathbf{s}} \|_{\hat{Z}}^2 \tag{13}$$

In order to practically find a solution of (13), we introduce the cost functional

$$\mathcal{J}(\mathbf{r}) = \frac{1}{2} \left\langle A\mathbf{r} - \tilde{\mathbf{s}}, \, A\mathbf{r} - \tilde{\mathbf{s}} \right\rangle_{\hat{Z}} \tag{14}$$

In the *basic approach* we propose to use the *gradient method* for finding the minimum of (14). In this method, in each iteration a correction $\delta\mathbf{r}$ is sought for a guess $\mathbf{r}$ which points into the negative gradient direction $-A^*(A\mathbf{r} - \tilde{\mathbf{s}})$ of the cost functional (14). In other words, starting with the initial guess $\mathbf{r}^{(0)} = 0$, the iteration of the gradient method goes as follows:

$$\mathbf{r}^{(0)} = 0$$
$$\mathbf{r}^{(n+1)} = \mathbf{r}^{(n)} - \beta^{(n)} A^*(A\mathbf{r}^{(n)} - \tilde{\mathbf{s}}), \tag{15}$$

where $\beta^{(n)}$ is the step-size at iteration number n. Notice that the signals $\mathbf{r}^{(n)}$ are measured at the base station, whereas the difference $A\mathbf{r}^{(n)} - \tilde{\mathbf{s}}$ is determined at the users. In particular, $\tilde{\mathbf{s}}$ is the pilot signal only known by the user who defined it, combined with zero signals at the remaining users. $A\mathbf{r}^{(n)}$ is the signal received by the users at the n-th iteration step.

9 The Adjoint Operator A^*

We see from (15) that we will have to apply the adjoint operator A^* repeatedly when implementing the gradient method. In this section we provide practically useful expressions for applying this operator to a given element of $\hat{Z}$.

First we mention that obviously $A^* = Q^*F^*\hat{M}^*$ due to the definition $A = \hat{M}FQ$.

Theorem 9.1. *We have*

$$M^* = \Gamma^{-1}Q, \quad \hat{M}^* = \Gamma^{-1}\hat{Q},$$
$$Q^* = M\Gamma, \quad \hat{Q}^* = \hat{M}\Gamma. \tag{16}$$

Proof: The proof is given in Appendix A.

Next, we want to find an expression for $F^*\mathbf{v}$, $\mathbf{v} \in U$, where F^* is the adjoint of the operator F.

Theorem 9.2. *Let $\mathbf{z}$ be the solution of the adjoint symmetric hyperbolic system*

$$-\Gamma(\mathbf{x})\frac{\partial \mathbf{z}}{\partial t} - \sum_{i=1}^{3} D^i \frac{\partial \mathbf{z}}{\partial x_i} + \Phi(\mathbf{x})\mathbf{z} = \Gamma(\mathbf{x})\mathbf{v}(\mathbf{x}, t), \tag{17}$$

$$\mathbf{z}(\mathbf{x}, T) = 0, \tag{18}$$
$$\mathbf{z}(\mathbf{x}, t) = 0 \quad \text{on} \quad \partial\Omega \times [0, T]. \tag{19}$$

Then

$$F^*\mathbf{v} = \Gamma^{-1}(\mathbf{x})\mathbf{z}(\mathbf{x}, t). \tag{20}$$

Proof: The proof is given in Appendix B.

Remark 9.1. This procedural characterization of the adjoint operator is often used in solution strategies of large scale inverse problems, where it naturally leads to so-called 'backpropagation strategies'. See for example [13, 14, 25, 27, 36, 48] and the references given there.

Remark 9.2. Notice that in the adjoint system (17)–(19) 'final value conditions' are given at $t = T$ in contrast to (1)–(4) where 'initial value conditions' are prescribed at $t = 0$. This corresponds to the fact that time is running backward in (17)–(19) and forward in (1)–(4).

10 The Acoustic Time-Reversal Mirror

In the following we want to define an operator $\mathcal{S}_a$ such that $F^* = \Gamma^{-1}\mathcal{S}_a F \mathcal{S}_a \Gamma$ holds. We will call this operator $\mathcal{S}_a$ the *acoustic time-reversal operator*. We will also define the *acoustic time-reversal mirrors* $\mathcal{T}_a$ and $\hat{\mathcal{T}}_a$, which act on the signals instead of the sources or fields.

We consider the acoustic system

$$\rho\frac{\partial \mathbf{v}_f}{\partial t} + \mathbf{grad}p_f(\mathbf{x}, t) = \mathbf{q_v}, \tag{21}$$

$$\kappa\frac{\partial p_f}{\partial t} + \mathrm{div}\mathbf{v}_f(\mathbf{x}, t) = \mathbf{q}_p, \tag{22}$$

$$\mathbf{v}_f(\mathbf{x}, 0) = 0, \qquad p_f(\mathbf{x}, 0) = 0 \quad \text{in} \quad \Omega \tag{23}$$

with $t \in [0, T]$ and zero boundary conditions at $\partial\Omega \times [0, T]$. We want to calculate the action of the *adjoint operator* F^* on a vector $(\phi, \psi)^T \in U$.

Theorem 10.1. *Let $(\phi, \psi)^T \in U$ and let $(\mathbf{v}_a, p_a)^T$ be the solution of the adjoint system*

$$-\rho\frac{\partial \mathbf{v}_a}{\partial t} - \mathbf{grad}p_a(\mathbf{x}, t) = \rho(\mathbf{x})\phi(\mathbf{x}, t) \tag{24}$$

$$-\kappa\frac{\partial p_a}{\partial t} - \mathrm{div}\mathbf{v}_a(\mathbf{x}, t) = \kappa(\mathbf{x})\psi(\mathbf{x}, t) \tag{25}$$

$$\mathbf{v}_a(\mathbf{x}, T) = 0, \qquad p_a(\mathbf{x}, T) = 0 \quad \text{in} \quad \Omega, \tag{26}$$

with $t \in [0, T]$ and zero boundary conditions at $\partial\Omega \times [0, T]$. Then we have

$$F^*\begin{pmatrix}\phi \\ \psi\end{pmatrix} = \begin{pmatrix}\rho^{-1}\mathbf{v}_a(\mathbf{x}, t) \\ \kappa^{-1}p_a(\mathbf{x}, t)\end{pmatrix}. \tag{27}$$

Proof: This theorem is just an application of Theorem 9.2 to the acoustic symmetric hyperbolic system. For the convenience of the reader, we will give a direct proof as well in Appendix C.

Definition 10.1. *We define the acoustic time-reversal operator $\mathcal{S}_a$ by putting for all $\mathbf{q} = (\mathbf{q_v}, \mathbf{q}_p)^T \in U$ (and similarly for all $\mathbf{u} = (\mathbf{v}, p)^T \in U$)*

$$(\mathcal{S}_a\mathbf{q})(\mathbf{x}, t) = \begin{pmatrix}-\mathbf{q_v}(\mathbf{x}, 2T - t) \\ \mathbf{q}_p(\mathbf{x}, 2T - t)\end{pmatrix} \tag{28}$$

Definition 10.2. *We define the acoustic time-reversal mirrors $\mathcal{T}_a$ and $\hat{\mathcal{T}}_a$ by putting for all $\mathbf{r} = (\mathbf{r_v}, \mathbf{r}_p)^T \in Z$ and all $\mathbf{s} = (\mathbf{s_v}, \mathbf{s}_p)^T \in \hat{Z}$*

$$(\mathcal{T}_a \mathbf{r})(t) = \begin{pmatrix} -\mathbf{r_v}(2T-t) \\ \mathbf{r}_p(2T-t) \end{pmatrix}, \quad (\hat{\mathcal{T}}_a \mathbf{s})(t) = \begin{pmatrix} -\mathbf{s_v}(2T-t) \\ \mathbf{s}_p(2T-t) \end{pmatrix} \tag{29}$$

The following lemma is easy to verify.

Lemma 10.1. *We have the following commutations*

$$\begin{aligned} M\mathcal{S}_a &= \mathcal{T}_a M, & \mathcal{S}_a Q &= Q\mathcal{T}_a, \\ \hat{M}\mathcal{S}_a &= \hat{\mathcal{T}}_a \hat{M}, & \mathcal{S}_a \hat{Q} &= \hat{Q}\hat{\mathcal{T}}_a. \end{aligned} \tag{30}$$

Theorem 10.2. *For $(\phi, \psi)^T \in U$ we have*

$$\Gamma^{-1}\mathcal{S}_a F \mathcal{S}_a \Gamma \begin{pmatrix} \phi \\ \psi \end{pmatrix} = F^* \begin{pmatrix} \phi \\ \psi \end{pmatrix} \tag{31}$$

Proof: For the proof it is convenient to make the following definition.

Definition 10.3. Acoustic time-reversal experiment: *For given $(\phi, \psi)^T \in U$ define $\mathbf{q_v}(\mathbf{x}, t) = \rho\phi$ and $\mathbf{q}_p(\mathbf{x}, t) = \kappa\psi$ and perform the following physical experiment*

$$\rho\frac{\partial}{\partial s}\mathbf{v}_{tr}(\mathbf{x}, s) + \mathbf{grad}p_{tr}(\mathbf{x}, s) = -\mathbf{q_v}(\mathbf{x}, 2T - s) \tag{32}$$

$$\kappa\frac{\partial}{\partial s}p_{tr}(\mathbf{x}, s) + \mathrm{div}\mathbf{v}_{tr}(\mathbf{x}, s) = q_p(\mathbf{x}, 2T - s) \tag{33}$$

$$\mathbf{v}_{tr}(\mathbf{x}, T) = 0, \qquad p_{tr}(\mathbf{x}, T) = 0 \quad in \quad \Omega, \tag{34}$$

with $s \in [T, 2T]$ and zero boundary conditions. Doing this experiment means to process the data in the following way: Time-reverse all data $(\phi, \psi)^T$ according to $t \to 2T - s$, $t \in [0, T]$, and, in addition, reverse the directions of the velocities $\phi \to -\phi$. We call this experiment the 'acoustic time-reversal experiment'. Notice that the time is running forward in this experiment.

The solution $(\mathbf{v}_{tr}, p_{tr})^T$ of this experiment can obviously be represented by

$$\begin{pmatrix} \mathbf{v}_{tr} \\ p_{tr} \end{pmatrix} = F\mathcal{S}_a\Gamma \begin{pmatrix} \phi \\ \psi \end{pmatrix}. \tag{35}$$

In order to show that the so defined time-reversal experiment is correctly modelled by the adjoint system derived above, we make the following change in variables:

$$\tau = 2T - s, \quad \hat{\mathbf{v}}_{tr} = -\mathbf{v}_{tr}, \quad \hat{p}_{tr} = p_{tr}, \tag{36}$$

which just corresponds to the application of the operator $\mathcal{S}_a$ to $(\mathbf{v}_{tr}, p_{tr})^T$. We have $\tau \in [0, T]$. In these variables the time-reversal system (32)–(34) gets the form

$$-\rho\frac{\partial}{\partial\tau}\hat{\mathbf{v}}_{tr}(\mathbf{x},\tau) - \mathbf{grad}\hat{p}_{tr}(\mathbf{x},\tau) = \mathbf{q_v}(\mathbf{x},\tau) \tag{37}$$

$$-\kappa\frac{\partial}{\partial\tau}\hat{p}_{tr}(\mathbf{x},\tau) - \mathrm{div}\hat{\mathbf{v}}_{tr}(\mathbf{x},\tau) = \mathbf{q}_p(\mathbf{x},\tau) \tag{38}$$

$$\hat{\mathbf{v}}_{tr}(\mathbf{x},T) = 0, \qquad \hat{p}_{tr}(\mathbf{x},T) = 0 \quad \text{in} \quad \Omega. \tag{39}$$

Taking into account the definition of $\mathbf{q_v}$ and $\mathbf{q}_p$, we see that

$$\begin{pmatrix} \hat{\mathbf{v}}_{tr} \\ \hat{p}_{tr} \end{pmatrix} = \begin{pmatrix} \mathbf{v}_a \\ p_a \end{pmatrix}$$

where $\mathbf{v}_a$ and p_a solve the adjoint system (24)–(26). Therefore, according to Theorem 10.1:

$$\begin{pmatrix} \hat{\mathbf{v}}_{tr} \\ \hat{p}_{tr} \end{pmatrix} = \Gamma F^* \begin{pmatrix} \phi \\ \psi \end{pmatrix}.$$

Since we have with (35), (36) also

$$\begin{pmatrix} \hat{\mathbf{v}}_{tr} \\ \hat{p}_{tr} \end{pmatrix} = \mathcal{S}_a F \mathcal{S}_a \Gamma \begin{pmatrix} \phi \\ \psi \end{pmatrix},$$

the theorem is proven.

11 The Electromagnetic Time-Reversal Mirror

In the following we want to define an operator $\mathcal{S}_e$ such that $F^* = \Gamma^{-1}\mathcal{S}_e F \mathcal{S}_e \Gamma$ holds. We will call this operator $\mathcal{S}_e$ the *electromagnetic time-reversal operator*. We will also define the *electromagnetic time-reversal mirrors* $\mathcal{T}_e$ and $\hat{\mathcal{T}}_e$, which act on the signals instead of the sources or fields.

We consider Maxwell's equations

$$\epsilon\frac{\partial\mathbf{E}_f}{\partial t} - \mathbf{curl}\,\mathbf{H}_f + \sigma\mathbf{E}_f = \mathbf{q}_E \tag{40}$$

$$\mu\frac{\partial\mathbf{H}_f}{\partial t} + \mathbf{curl}\,\mathbf{E}_f = \mathbf{q}_H \tag{41}$$

$$\mathbf{E}_f(\mathbf{x},0) = 0, \qquad \mathbf{H}_f(\mathbf{x},0) = 0 \tag{42}$$

with $t \in [0,T]$ and zero boundary conditions. We want to calculate the action of the adjoint operator F^* on a vector $(\phi,\psi)^T \in U$.

Theorem 11.1. *Let $(\phi,\psi)^T \in U$ and let $(\mathbf{E}_a,\mathbf{H}_a)^T$ be the solution of the adjoint system*

$$-\epsilon\frac{\partial\mathbf{E}_a}{\partial t} + \mathbf{curl}\mathbf{H}_a(\mathbf{x},t) + \sigma\mathbf{E}_a = \epsilon(\mathbf{x})\phi(\mathbf{x},t) \tag{43}$$

$$-\mu\frac{\partial\mathbf{H}_a}{\partial t} - \mathbf{curl}\mathbf{E}_a(\mathbf{x},t) = \mu(\mathbf{x})\psi(\mathbf{x},t) \tag{44}$$

$$\mathbf{E}_a(\mathbf{x}, T) = 0, \qquad \mathbf{H}_a(\mathbf{x}, T) = 0 \quad \text{in} \quad \Omega \tag{45}$$

with $t \in [0, T]$ and zero boundary conditions. Then we have

$$F^* \begin{pmatrix} \phi \\ \psi \end{pmatrix} = \begin{pmatrix} \epsilon^{-1} \mathbf{E}_a(\mathbf{x}, t) \\ \mu^{-1} \mathbf{H}_a(\mathbf{x}, t) \end{pmatrix}. \tag{46}$$

Proof: This theorem is just an application of Theorem 9.2 to the electromagnetic symmetric hyperbolic system. Again, we will give a direct proof in Appendix D as well.

Definition 11.1. We define the **electromagnetic time-reversal operator** $\mathcal{S}_e$ by putting for all $\mathbf{q} = (\mathbf{q}_E, \mathbf{q}_H)^T \in U$ (and similarly for all $\mathbf{u} = (\mathbf{E}, \mathbf{H})^T \in U$)

$$(\mathcal{S}_e \mathbf{q})(\mathbf{x}, t) = \begin{pmatrix} -\mathbf{q}_E(\mathbf{x}, 2T - t) \\ \mathbf{q}_H(\mathbf{x}, 2T - t) \end{pmatrix} \tag{47}$$

Definition 11.2. We define the **electromagnetic time-reversal mirrors** $\mathcal{T}_e$ and $\hat{\mathcal{T}}_e$ by putting for all $\mathbf{r} = (\mathbf{r}_E, \mathbf{r}_H)^T \in Z$ and for all $\mathbf{s} = (\mathbf{s}_E, \mathbf{s}_H)^T \in \hat{Z}$

$$(\mathcal{T}_e \mathbf{r})(t) = \begin{pmatrix} -\mathbf{r}_E(2T - t) \\ \mathbf{r}_H(2T - t) \end{pmatrix}, \quad (\hat{\mathcal{T}}_e \mathbf{s})(t) = \begin{pmatrix} -\mathbf{s}_E(2T - t) \\ \mathbf{s}_H(2T - t) \end{pmatrix} \tag{48}$$

The following lemma is easy to verify.

Lemma 11.1. We have the following commutations

$$\begin{align} M\mathcal{S}_e &= \mathcal{T}_e M, & \mathcal{S}_e Q &= Q\mathcal{T}_e, \\ \hat{M}\mathcal{S}_e &= \hat{\mathcal{T}}_e \hat{M}, & \mathcal{S}_e \hat{Q} &= \hat{Q}\hat{\mathcal{T}}_e. \end{align} \tag{49}$$

Theorem 11.2. For $(\phi, \psi)^T \in U$ we have

$$\Gamma^{-1} \mathcal{S}_e F \mathcal{S}_e \Gamma \begin{pmatrix} \phi \\ \psi \end{pmatrix} = F^* \begin{pmatrix} \phi \\ \psi \end{pmatrix} \tag{50}$$

Proof: For the proof it is convenient to make the following definition.

Definition 11.3. Electromagnetic time-reversal experiment: For a given vector $(\phi, \psi)^T \in U$ define $\mathbf{q}_E(\mathbf{x}, t) = \epsilon\phi$ and $\mathbf{q}_H(\mathbf{x}, t) = \mu\psi$ and perform the physical experiment

$$\epsilon \frac{\partial}{\partial s} \mathbf{E}_{tr}(\mathbf{x}, s) - \mathbf{curl} \mathbf{H}_{tr}(\mathbf{x}, s) + \sigma \mathbf{E}_{tr}(\mathbf{x}, s) = -\mathbf{q}_E(\mathbf{x}, 2T - s) \tag{51}$$

$$\mu \frac{\partial}{\partial s} \mathbf{H}_{tr}(\mathbf{x}, s) + \mathbf{curl} \mathbf{E}_{tr}(\mathbf{x}, s) = \mathbf{q}_H(\mathbf{x}, 2T - s) \tag{52}$$

$$\mathbf{E}_{tr}(\mathbf{x}, T) = 0, \qquad \mathbf{H}_{tr}(\mathbf{x}, T) = 0 \quad \text{in} \quad \Omega, \tag{53}$$

with $s \in [T, 2T]$ and zero boundary conditions. Doing this experiment means to process the data in the following way: Time-reverse all data according to $t \to 2T - s$, $t \in [0, T]$, and, in addition, reverse the directions of the electric field component by $\phi \to -\phi$. We call this experiment the 'electromagnetic time-reversal experiment'. Notice that the time is running forward in this experiment.

The solution $(\mathbf{E}_{tr}, \mathbf{H}_{tr})^T$ of this experiment can obviously be represented by

$$\begin{pmatrix} \mathbf{E}_{tr} \\ \mathbf{H}_{tr} \end{pmatrix} = F \mathcal{S}_e \Gamma \begin{pmatrix} \phi \\ \psi \end{pmatrix}. \tag{54}$$

In order to show that the so defined time-reversal experiment is correctly modelled by the adjoint system derived above, we make the following change in variables:

$$\tau = 2T - s, \quad \hat{\mathbf{E}}_{tr} = -\mathbf{E}_{tr}, \quad \hat{\mathbf{H}}_{tr} = \mathbf{H}_{tr}, \tag{55}$$

which just corresponds to the application of the operator $\mathcal{S}_e$ to $(\mathbf{E}_{tr}, \mathbf{H}_{tr})^T$. We have $\tau \in [0, T]$. In these variables the time-reversal system (51)–(53) gets the form

$$-\epsilon \frac{\partial}{\partial \tau} \hat{\mathbf{E}}_{tr}(\mathbf{x}, \tau) + \mathbf{curl}\hat{\mathbf{H}}_{tr}(\mathbf{x}, \tau) + \sigma \hat{\mathbf{E}}_{tr}(\mathbf{x}, s) = \mathbf{q}_\mathbf{E}(\mathbf{x}, \tau), \tag{56}$$

$$-\mu \frac{\partial}{\partial \tau} \hat{\mathbf{H}}_{tr}(\mathbf{x}, \tau) - \mathbf{curl}\hat{\mathbf{E}}_{tr}(\mathbf{x}, \tau) = \mathbf{q}_\mathbf{H}(\mathbf{x}, \tau) \tag{57}$$

$$\hat{\mathbf{E}}_{tr}(\mathbf{x}, T) = 0, \qquad \hat{\mathbf{H}}_{tr}(\mathbf{x}, T) = 0 \quad \text{in} \quad \Omega. \tag{58}$$

Taking into account the definition of $\mathbf{q}_E$ and $\mathbf{q}_H$, we see that

$$\begin{pmatrix} \hat{\mathbf{E}}_{tr} \\ \hat{\mathbf{H}}_{tr} \end{pmatrix} = \begin{pmatrix} \mathbf{E}_a \\ \mathbf{H}_a \end{pmatrix}$$

where $\mathbf{E}_e$ and $\mathbf{H}_e$ solve the adjoint system (43)–(45). Therefore, according to Theorem 11.1:

$$\begin{pmatrix} \hat{\mathbf{E}}_{tr} \\ \hat{\mathbf{H}}_{tr} \end{pmatrix} = \Gamma F^* \begin{pmatrix} \phi \\ \psi \end{pmatrix}.$$

Since we have with (54), (55) also

$$\begin{pmatrix} \hat{\mathbf{E}}_{tr} \\ \hat{\mathbf{H}}_{tr} \end{pmatrix} = \mathcal{S}_e F \mathcal{S}_e \Gamma \begin{pmatrix} \phi \\ \psi \end{pmatrix},$$

the theorem is proven.

Remark 11.1. For electromagnetic waves, there is formally an alternative way to define the *electromagnetic time-reversal operator*, namely putting for all $\mathbf{q} = (\mathbf{q}_E, \mathbf{q}_H)^T \in U$

$$(\mathcal{S}_e \mathbf{q})(\mathbf{x}, t) = \begin{pmatrix} \mathbf{q}_E(\mathbf{x}, 2T - t) \\ -\mathbf{q}_H(\mathbf{x}, 2T - t) \end{pmatrix},$$

accompanied by the analogous definitions for the *electromagnetic time-reversal mirrors*. With these alternative definitions, Theorem 11.2 holds true as well, with only very few changes in the proof. Which form to use depends mainly on the preferred form for modelling applied antenna signals in the given antenna system. The first formulation directly works with applied electric currents, whereas the second form is useful for example for magnetic dipole sources.

12 Time-Reversal and the Adjoint Operator A^*

Define $\mathcal{S} = \mathcal{S}_a$, $\mathcal{T} = \mathcal{T}_a$, $\hat{\mathcal{T}} = \hat{\mathcal{T}}_a$ for the acoustic case, and $\mathcal{S} = \mathcal{S}_e$, $\mathcal{T} = \mathcal{T}_e$, $\hat{\mathcal{T}} = \hat{\mathcal{T}}_e$ for the electromagnetic case. We call $\mathcal{S}$ the *time-reversal operator* and $\mathcal{T}$, $\hat{\mathcal{T}}$ the *time-reversal mirrors*. We combine the results of Lemma 10.1 and Lemma 11.1 into the following lemma.

Lemma 12.1. *We have the following commutations*

$$MS = TM, \qquad SQ = QT,$$
$$\hat{M}S = \hat{T}\hat{M}, \qquad S\hat{Q} = \hat{Q}\hat{T}. \tag{59}$$

Moreover, combining Theorem 10.2 and Theorem 11.2 we get

Theorem 12.1. *For $(\phi, \psi)^T \in U$ we have*

$$\Gamma^{-1}SFS\Gamma \begin{pmatrix} \phi \\ \psi \end{pmatrix} = F^* \begin{pmatrix} \phi \\ \psi \end{pmatrix}. \tag{60}$$

With this, we can prove the following theorem which provides the fundamental link between time-reversal and inverse problems.

Theorem 12.2. *We have*

$$A^* = \mathcal{T}B\hat{\mathcal{T}}. \tag{61}$$

Proof: Recall that the adjoint operator A^* can be decomposed as $A^* = Q^*F^*\hat{M}^*$. With Theorem 9.1, Theorem 12.1, and Lemma 12.1, it follows therefore that

$$A^* = M\Gamma\Gamma^{-1}SFS\Gamma\Gamma^{-1}\hat{Q}$$
$$= MSFS\hat{Q}$$
$$= \mathcal{T}MF\hat{Q}\hat{\mathcal{T}}$$
$$= \mathcal{T}B\hat{\mathcal{T}},$$

which proves the theorem.

Remark 12.1. The above theorem provides a direct link between the adjoint operator A^*, which plays a central role in the theory of inverse problems, and a physical experiment modelled by B. The expression $\mathcal{T}B\hat{\mathcal{T}}$ defines a 'time-reversal experiment'. We will demonstrate in the following sections how we can make use of this relationship in order to solve the inverse problem of communication by a series of physical time-reversal experiments.

Remark 12.2. We mention that the above results hold as well for *elastic waves* with a suitable definition of the *elastic time-reversal mirrors*. We leave out the details for brevity.

13 Iterative Time-Reversal for the Gradient Method

13.1 The Basic Version

The results achieved above give rise to the following experimental procedure for applying the gradient method (15) to the inverse problem of communication as formulated in Sects. 7 and 8. First, the pilot signal $\tilde{s}(t)$ is defined by user $\mathcal{U}_1$ as described in (11). Moreover, we assume that the first guess $\mathbf{r}^{(0)}(t)$ at the base station is chosen to be zero. Then, using Theorem 12.2, we can write the gradient method (15) in the equivalent form

$$\mathbf{r}^{(0)} = 0$$
$$\mathbf{r}^{(n+1)} = \mathbf{r}^{(n)} + \beta^{(n)} \mathcal{T} B \hat{\mathcal{T}} (\tilde{s} - A\mathbf{r}^{(n)}), \tag{62}$$

or, expanding it,

$$\mathbf{r}^{(0)} = 0$$
$$\mathbf{s}^{(n)} = \tilde{s} - A\mathbf{r}^{(n)} \tag{63}$$
$$\mathbf{r}^{(n+1)} = \mathbf{r}^{(n)} + \beta^{(n)} \mathcal{T} B \hat{\mathcal{T}} \mathbf{s}^{(n)}$$

In a more detailed form, we arrive at the following *experimental procedure for implementing the gradient method*, where we fix in this description $\beta^{(n)} = 1$ for all iteration numbers n for simplicity:

1. The user $\mathcal{U}_1$ chooses a pilot signal $\alpha(t)$ to be used for communicating with the base station. The objective signal at all users is then $\tilde{s}(t) = (\alpha(t), 0, \dots, 0)^T$. The initial guess $\mathbf{r}^{(0)}(t)$ at the base station is defined to be zero, such that $\mathbf{s}^{(0)} = \tilde{s}$.
2. The user $\mathcal{U}_1$ initiates the communication by sending the time-reversed pilot signal into the environment. This signal is $\hat{\mathcal{T}} \mathbf{s}^{(0)}(t)$. All other users are quiet.
3. The base station receives the pilot signal as $B\hat{\mathcal{T}} \mathbf{s}^{(0)}(t)$. It time-reverses this signal and sends this time-reversed form, namely $\mathbf{r}^{(1)}(t) = \mathcal{T} B \hat{\mathcal{T}} \mathbf{s}^{(0)}(t)$, back into the medium.
4. The new signal arrives at all users as $A\mathbf{r}^{(1)}(t)$. All users compare the received signals with their components of the objective signal $\tilde{s}(t)$. They take the difference $\mathbf{s}^{(1)}(t) = \tilde{s}(t) - A\mathbf{r}^{(1)}(t)$, and time-reverse it. They send this new signal $\hat{\mathcal{T}} \mathbf{s}^{(1)}(t)$ back into the medium.
5. The base station receives this new signal, time-reverses it, adds it to the previous signal $\mathbf{r}^{(1)}(t)$, and sends the sum back into the medium as $\mathbf{r}^{(2)}(t)$.
6. This iteration is continued until all users are satisfied with the match between the received signal $A\mathbf{r}^{(n)}(t)$ and the objective signal $\tilde{s}(t)$ at some iteration number n. Alternatively, a fixed iteration number n can be specified a-priori for stopping the iteration.

Needless to say that, in practical implementations, the laboratory time needs to be reset to zero after each time-reversal step.

Remark 13.1. The experimental procedure which is described above is practically equivalent to the experimental procedure which was suggested and experimentally verified in [34, 35]. Therefore, our basic scheme provides an alternative derivation and interpretation of this experimental procedure.

Remark 13.2. We mention that several refinements of this scheme are possible and straightforward. For example, a weighted inner product can be introduced for the user signal space $\hat{Z}$ which puts different preferences on the satisfaction of the user objectives during the iterative optimization process. For example, if the 'importance' of suppressing interferences with other users is valued higher than to get an optimal signal quality at the specified user $\mathcal{U}_1$, a higher weight can be put into the inner product at those users which did not start the communication process. A user who does not care about these interferences, simply puts a very small weight into his component of the inner product of $\hat{Z}$.

Remark 13.3. Notice that there is no mechanism directly built into this procedure which prevents the energy emitted by the base antenna to increase more than the communication system can support. For example, if the subspace of signals

$$Z_0 := \{\mathbf{r}(t) \,:\, A\mathbf{r} = 0\}$$

is not empty, then it might happen that during the iteration described above (e.g. due to noise) an increasing amount of energy is put into signals emitted by the base station which are in this subspace and which all produce zero contributions to the measurements at all users. More generally, elements of the subspace of signals

$$Z_\varepsilon := \{\mathbf{r}(t) \,:\, \|A\mathbf{r}\|_{\hat{Z}} < \varepsilon\|\mathbf{r}\|_Z\},$$

for a very small threshold $0 < \epsilon << 1$, might cause problems during the iteration if the pilot signal $\tilde{\mathbf{s}}(t)$ chosen by the user has contributions in the subspace AZ_ε (i.e. in the space of all $\mathbf{s} = A\mathbf{r}$ with $\mathbf{r} \in Z_\varepsilon$). This is so because in the effort of decreasing the mismatch between $A\mathbf{r}$ and $\tilde{\mathbf{s}}(t)$, the base antenna might need to put signals with high energy into the system in order to get only small improvements in the signal match at the user side. Since the environment (and therefore the operator A) is unknown a-priori, it is difficult to avoid the existence of such contributions in the pilot signal.

One possible way to prevent the energy emitted by the base station to increase artificially would be to project the signals $\mathbf{r}^{(n)}(t)$ onto the orthogonal complements of the subspaces Z_0 or Z_ε (if they are known or can be constructed by some means) prior to their emission. Alternatively, the iteration can be stopped at an early stage before these unwanted contributions start to build up. (This in fact has been suggested in [34, 35]).

In the following subsection we introduce an alternative way of ensuring that the energy emitted by the base station stays reasonably bounded in the effort of fitting the pilot signal at the users.

13.2 The Regularized Version

Consider the regularized problem

$$\mathbf{r}_{LSr} = \text{Min}_{\mathbf{r}} \left(\|A\mathbf{r} - \tilde{\mathbf{s}}\|_{\hat{Z}}^2 + \lambda \|\mathbf{r}\|_Z^2 \right) \tag{64}$$

with some suitably chosen regularization parameter $\lambda > 0$. In this problem formulation a trade-off is sought between a signal fit at the user side and a minimized energy emission at the base station. The trade-off parameter is the regularization parameter λ. Instead of (14) we need to consider now

$$\tilde{\mathcal{J}}(\mathbf{r}) = \frac{1}{2} \Big\langle A\mathbf{r} - \tilde{\mathbf{s}}, A\mathbf{r} - \tilde{\mathbf{s}} \Big\rangle_{\hat{Z}} + \frac{\lambda}{2} \Big\langle \mathbf{r}, \mathbf{r} \Big\rangle_Z \tag{65}$$

The negative gradient direction is now given by $-A^*(A\mathbf{r} - \tilde{\mathbf{s}}) - \lambda \mathbf{r}$, such that the regularized iteration reads:

$$\begin{aligned} \mathbf{r}^{(0)} &= 0 \\ \mathbf{r}^{(n+1)} &= \mathbf{r}^{(n)} + \beta^{(n)} \Big(\mathcal{T} B \hat{\mathcal{T}} (\tilde{\mathbf{s}} - A\mathbf{r}^{(n)}) - \lambda \mathbf{r}^{(n)} \Big), \end{aligned} \tag{66}$$

where we have replaced A^* by $\mathcal{T} B \hat{\mathcal{T}}$. The time-reversal iteration can be expanded into the following practical scheme

$$\begin{aligned} \mathbf{r}^{(0)} &= 0 \\ \mathbf{s}^{(n)} &= \tilde{\mathbf{s}} - A\mathbf{r}^{(n)} \\ \mathbf{r}^{(n+1)} &= \mathbf{r}^{(n)} + \beta^{(n)} \mathcal{T} B \hat{\mathcal{T}} \mathbf{s}^{(n)} - \beta^{(n)} \lambda \mathbf{r}^{(n)}. \end{aligned} \tag{67}$$

Comparing with (63), we see that the adaptations which need to be applied in the practical implementation for stabilizing the basic algorithm can easily be done.

14 The Minimum Norm Solution Approach

14.1 The Basic Version

In this section we want to propose an alternative scheme for solving the inverse problem of communication. As mentioned above, a major drawback of the basic approach (13) is that the energy emitted by the base station is not limited explicitly when solving the optimization problem. The regularized version presented above alleviates this problem. However, we want to mention here that, under certain assumptions, there is an alternative scheme which can be employed instead and which has an energy constraint directly built in. Under the formal assumption that there exists at least one (and presumably more than one) solution of the inverse problem at hand (i.e. the 'formally underdetermined case'), we can look for the *minimum norm solution*

$$\text{Min}_{\mathbf{r}} \|\mathbf{r}\|_Z \quad \text{subject to} \quad A\mathbf{r} = \tilde{\mathbf{s}}. \tag{68}$$

In Hilbert spaces this solution has an explicit form. It is

$$\mathbf{r}_{MN} = A^*(AA^*)^{-1}\tilde{\mathbf{s}}. \tag{69}$$

Here, the operator $(AA^*)^{-1}$ acts as a filter on the pilot signal $\tilde{\mathbf{s}}$. Instead of sending the pilot signal to the base station, the users send the filtered version of it. Certainly, a method must be found in order to apply the filter $(AA^*)^{-1}$ to the pilot signal. One possibility of doing so would be to try to determine the operator AA^* explicitly by a series of time-reversal experiments on some set of basis functions of $\hat{Z}$, and then invert this operator numerically. However, this might not be practical in many situations. It certainly would be slow and it would involve a significant amount of signal-processing. Therefore, we propose an alternative procedure. First, we notice that there is no need to determine the whole operator $(AA^*)^{-1}$, but that we only have to apply it to one specific signal, namely $\tilde{\mathbf{s}}$. Let us introduce the short notation

$$C = AA^*.$$

In this notation, we are looking for a signal $\hat{\mathbf{s}} \in \hat{Z}$ such that $C\hat{\mathbf{s}} = \tilde{\mathbf{s}}$. We propose to solve this equation in the least squares sense:

$$\hat{\mathbf{s}} = \text{Min}_{\mathbf{s}} \|C\mathbf{s} - \tilde{\mathbf{s}}\|_{\hat{Z}}^2. \tag{70}$$

Moreover, as a suitable method for practically finding this solution, we want to use the *gradient method*. Starting with the initial guess $\mathbf{s}_0 = 0$, the gradient method reads

$$\mathbf{s}^{(0)} = 0$$
$$\mathbf{s}^{(n+1)} = \mathbf{s}^{(n)} - \beta^{(n)} C^*(C\mathbf{s}^{(n)} - \tilde{\mathbf{s}}), \tag{71}$$

where C^* is the adjoint operator to C and $\beta^{(n)}$ is again some step-size. Expanding this expression, and taking into account $C^* = C$ and $A^* = \mathcal{T}B\hat{\mathcal{T}}$, we arrive at

$$\mathbf{s}^{(0)} = 0$$
$$\mathbf{s}^{(n+1)} = \mathbf{s}^{(n)} + \beta^{(n)} A\mathcal{T}B\hat{\mathcal{T}}(\tilde{\mathbf{s}} - A\mathcal{T}B\hat{\mathcal{T}}\mathbf{s}^{(n)}). \tag{72}$$

In the practical implementation, we arrive at the following iterative scheme:

$$\text{Initialize gradient iteration:} \quad \mathbf{s}^{(0)} = 0$$

$$\text{for } n = 0, 1, 2, \ldots \text{ do:}$$

$$\mathbf{r}^{(n+\frac{1}{2})} = \mathcal{T}B\hat{\mathcal{T}}\mathbf{s}^{(n)}$$

$$\mathbf{s}^{(n+\frac{1}{2})} = \tilde{\mathbf{s}} - A\mathbf{r}^{(n+\frac{1}{2})}$$

$$\mathbf{r}^{(n+1)} = \mathcal{T}B\hat{\mathcal{T}}\mathbf{s}^{(n+\frac{1}{2})} \tag{73}$$

$$\mathbf{s}^{(n+1)} = \mathbf{s}^{(n)} + \beta^{(n)}A\mathbf{r}^{(n+1)}$$

$$\text{Terminate iteration at step } \hat{n}: \quad \hat{\mathbf{s}} = \mathbf{s}^{(\hat{n})}$$

$$\text{Final result:} \quad \mathbf{r}_{MN} = \mathcal{T}B\hat{\mathcal{T}}\hat{\mathbf{s}}.$$

This iteration can be implemented by a series of time-reversal experiments, without the need of heavy signal-processing. The final step of the above algorithm amounts to applying A^* to the result of the gradient iteration for calculating $\hat{\mathbf{s}} = (AA^*)^{-1}\tilde{\mathbf{s}}$, which yields then $\mathbf{r}_{MN}$. This will then be the signal to be applied by the base station during the communication process with the user $\mathcal{U}_1$.

14.2 The Regularized Version

In some situations it might be expected that the operator C is ill-conditioned, such that its inversion might cause instabilities, in particular when noisy signals are involved. For those situations, a regularized form of the minimum norm solution is available, namely

$$\mathbf{r}_{MNr} = A^*(AA^* + \lambda I_{\hat{Z}})^{-1}\tilde{\mathbf{s}} \tag{74}$$

where $I_{\hat{Z}}$ denotes the identity operator in $\hat{Z}$ and $\lambda > 0$ is some suitably chosen regularization parameter. The necessary adjustments in the gradient iteration for applying $(AA^* + \lambda I_{\hat{Z}})^{-1}$ to $\tilde{\mathbf{s}}$ are easily done. We only mention here the resulting procedure for the implementation of this gradient method by a series of time-reversal experiments:

$$\text{Initialize gradient iteration:} \quad \mathbf{s}^{(0)} = 0$$

$$\text{for } n = 0, 1, 2, \ldots \text{ do:}$$

$$\mathbf{r}^{(n+\frac{1}{2})} = \mathcal{T}B\hat{\mathcal{T}}\mathbf{s}^{(n)}$$

$$\mathbf{s}^{(n+\frac{1}{2})} = \tilde{\mathbf{s}} - A\mathbf{r}^{(n+\frac{1}{2})} - \lambda\mathbf{s}^{(n)}$$

$$\mathbf{r}^{(n+1)} = \mathcal{T}B\hat{\mathcal{T}}\mathbf{s}^{(n+\frac{1}{2})} \tag{75}$$

$$\mathbf{s}^{(n+1)} = \mathbf{s}^{(n)} + \beta^{(n)}A\mathbf{r}^{(n+1)} + \beta^{(n)}\lambda\mathbf{s}^{(n+\frac{1}{2})}.$$

$$\text{Terminate iteration at step } \hat{n}: \quad \hat{\mathbf{s}} = \mathbf{s}^{(\hat{n})}$$

$$\text{Final result:} \quad \mathbf{r}_{MNr} = \mathcal{T}B\hat{\mathcal{T}}\hat{\mathbf{s}}.$$

Again, the last step shown above is a final application of A^* to the result of the gradient iteration for calculating $\hat{\mathbf{s}} = (AA^* + \lambda I_{\hat{Z}})^{-1}\tilde{\mathbf{s}}$, which yields then $\mathbf{r}_{MNr}$. This will then be the signal to be applied by the base station during the communication process with the user $\mathcal{U}_1$.

15 The Regularized Least Squares Solution Revisited

We have introduced above the regularized least squares solution of the inverse problem of communication, namely

$$\mathbf{r}_{LSr} = \text{Min}_{\mathbf{r}}\left(\|A\mathbf{r} - \tilde{\mathbf{s}}\|_{\hat{Z}}^2 + \lambda\|\mathbf{r}\|_Z^2\right) \tag{76}$$

with $\lambda > 0$ being the regularization parameter. In Hilbert spaces, the solution of (76) has an explicit form. It is

$$\mathbf{r}_{LSr} = (A^*A + \lambda I_Z)^{-1}A^*\tilde{\mathbf{s}}, \tag{77}$$

where I_Z is the identity operator in Z. It is therefore tempting to try to implement also this direct form as a series of time-reversal experiments and compare its performance with the gradient method as it was described above. As our last strategy which we present in these lecture notes we want to show that such an alternative direct implementation of (76) is in fact possible.

Notice that in (76) the filtering operator $(A^*A + \lambda I_Z)^{-1}$ is applied at the base station, in contrast to the previous case where the user signal was filtered by the operator $(AA^* + \lambda I_{\hat{Z}})^{-1}$. Analogously to the previous case, we need to find a practical way to apply this filter to a signal at the base station. We propose again to solve the equation

$$(A^*A + \lambda I_Z)\hat{\mathbf{r}} = \tilde{\mathbf{r}} \tag{78}$$

in the least squares sense, where $\tilde{\mathbf{r}} = A^*\tilde{\mathbf{s}}$. Defining

$$C = A^*A + \lambda I_Z,$$

and using $C^* = C$ and $A^* = \mathcal{T}B\hat{\mathcal{T}}$, we arrive at the following *gradient iteration* for solving problem (78):

$$\begin{aligned} \mathbf{r}^{(0)} &= 0 \\ \mathbf{r}^{(n+1)} &= \mathbf{r}^{(n)} + \beta^{(n)}(\mathcal{T}B\hat{\mathcal{T}}A + \lambda I_Z)\left(\tilde{\mathbf{r}} - (\mathcal{T}B\hat{\mathcal{T}}A + \lambda I_Z)\mathbf{r}^{(n)}\right). \end{aligned} \tag{79}$$

This gives rise to the following practical implementation by a series of time-reversal experiments:

User sends pilot signal to base station: $\tilde{\mathbf{r}} = T\hat{T}\tilde{\mathbf{s}}$

Initialize gradient iteration: $\mathbf{r}^{(0)} = 0$

for $n = 0, 1, 2, \ldots$ do:

$$\mathbf{s}^{(n+\frac{1}{2})} = A\mathbf{r}^{(n)}$$

$$\mathbf{r}^{(n+\frac{1}{2})} = \tilde{\mathbf{r}} - T B \hat{T} \mathbf{s}^{(n+\frac{1}{2})} - \lambda \mathbf{r}^{(n)}$$

$$\mathbf{s}^{(n+1)} = A\mathbf{r}^{(n+\frac{1}{2})}$$

$$\mathbf{r}^{(n+1)} = \mathbf{r}^{(n)} + \beta^{(n)} T B \hat{T} \mathbf{s}^{(n+1)} + \beta^{(n)} \lambda \mathbf{r}^{(n+\frac{1}{2})}.$$

Terminate iteration at step $\hat{n}$: $\hat{\mathbf{r}} = \mathbf{r}^{(\hat{n})}$,

Final result: $\mathbf{r}_{LSr} = \hat{\mathbf{r}}$.

(80)

$\mathbf{r}_{LSr}$ will then be the signal to be applied by the base station during the communication process with the user $\mathcal{U}_1$.

16 Partial and Generalized Measurements

In many practical applications, only partial measurements of the whole wave-field are available. For example, in ocean acoustics often only pressure is measured, whereas the velocity field is not part of the measurement process. Similarly, in wireless communication only one or two components of the electric field might be measured simultaneously, but the remaining electric components and all magnetic components are missing. We want to demonstrate in this section that all results presented above are valid also in this situation of partial measurements, with the suitable adaptations.

Mathematically, the measurement operator needs to be adapted for the situation of partial measurements. Let us concentrate here on the special situation that only one component $\mathbf{u}_\nu$ ($\nu \in \{1, 2, 3, \ldots\}$) of the incoming wave field $\mathbf{u}$ is measured by the users and the base station. All other possible situations will then just be combinations of this particular case. It might also occur the situation that users can measure a different partial set of components than the base station. That case also follows directly from this canonical situation.

We introduce the new signal space at the base station $Y = (L_2[0, T])^K$ and the corresponding 'signal projection operator' P_ν by putting

$$P_\nu : Z \to Y, \quad P_\nu(\mathbf{r}_1(t), \ldots, \mathbf{r}_K(t))^T = \mathbf{r}_\nu(t).$$

We see immediately that its adjoint P_ν^* is given by

$$P_\nu^* : Y \to Z, \quad P_\nu^* \mathbf{r}_\nu(t) = (0, \ldots, 0, \mathbf{r}_\nu(t), 0, \ldots, 0)^T$$

where $\mathbf{r}_\nu(t)$ appears on the right hand side at the ν-th position. Our new measurement operator M_ν, and the new source operator Q_ν, are then defined by

$$M_\nu : U \to Y, \qquad M_\nu \mathbf{u} = P_\nu M \mathbf{u}$$
$$Q_\nu : Y \to U, \qquad Q_\nu \mathbf{r}_\nu = Q P_\nu^* \mathbf{r}_\nu. \tag{81}$$

Analogous definitions are done for $\hat{Y}$, $\hat{P}_\nu$, $\hat{M}_\nu$ and $\hat{Q}_\nu$ at the users.

Obviously, we will have to replace now in the above derivation of the iterative time-reversal procedure all measurement operators M by M_ν (and $\hat{M}$ by $\hat{M}_\nu$) and all source operators Q by Q_ν (and $\hat{Q}$ by $\hat{Q}_\nu$). In particular, the new 'communication operators' are now given by

$$A_\nu : Y \to \hat{Y}, \quad A_\nu \mathbf{r}_\nu = \hat{M}_\nu F Q_\nu \mathbf{r}_\nu,$$
$$B_\nu : \hat{Y} \to Y, \quad B_\nu \mathbf{s}_\nu = M_\nu F \hat{Q}_\nu \mathbf{s}_\nu. \tag{82}$$

In the following two theorems we show that the main results presented so far carry over to these newly defined operators.

Theorem 16.1. *We have*

$$M_\nu^* = \Gamma^{-1} Q_\nu, \quad \hat{M}_\nu^* = \Gamma^{-1} \hat{Q}_\nu,$$
$$Q_\nu^* = M_\nu \Gamma, \qquad \hat{Q}_\nu^* = \hat{M}_\nu \Gamma. \tag{83}$$

Proof: The proof is an easy exercise using (81) and Theorem 9.1.

Theorem 16.2. *It is*

$$A_\nu^* = \mathcal{T} B_\nu \hat{\mathcal{T}}. \tag{84}$$

Proof: The proof is now identical to the proof of Theorem 12.2, using Theorem 16.1 instead of Theorem 9.1.

Remark 16.1. In fact, it is easy to verify that all results of these notes remain valid for *arbitrarily defined linear measurement operators*

$$M_\mathcal{U} : U \to Z_\mathcal{U}, \qquad M_\mathcal{A} : U \to Z_\mathcal{A},$$

where $Z_\mathcal{U}$ and $Z_\mathcal{A}$ are any meaningful signal spaces at the users and the base antennas, respectively. The only requirement is that it is experimentally possible to apply signals according to the source operators defined by

$$Q_\mathcal{U} : Z_\mathcal{U} \to U, \qquad Q_\mathcal{U} = \Gamma M_\mathcal{U}^*$$
$$Q_\mathcal{A} : Z_\mathcal{A} \to U, \qquad Q_\mathcal{A} = \Gamma M_\mathcal{A}^*$$

where $M_\mathcal{U}^*$ and $M_\mathcal{A}^*$ are the formal adjoint operators to $M_\mathcal{U}$ and $M_\mathcal{A}$ with respect to the chosen signal spaces $Z_\mathcal{U}$ and $Z_\mathcal{A}$. In addition, the measurement and source operators as defined above are required to satisfy the commutation relations as stated in Lemma 12.1. Under these assumptions, we define the *generalized communication operators* $\tilde{A}$ and $\tilde{B}$ by

$$\tilde{A} : Z_{\mathcal{A}} \to Z_{\mathcal{U}}, \quad \tilde{A}\mathbf{r} = M_{\mathcal{U}}FQ_{\mathcal{A}}\mathbf{r},$$
$$\tilde{B} : Z_{\mathcal{U}} \to Z_{\mathcal{A}}, \quad \tilde{B}\mathbf{s} = M_{\mathcal{A}}FQ_{\mathcal{U}}\mathbf{s}.$$

Now the proof to Theorem 12.2 directly carries over to this generalized situation, such that we have also here

$$\tilde{A}^* = \mathcal{T}\tilde{B}\hat{\mathcal{T}}.$$

This yields iterative time-reversal schemes completely analogous to those presented above.

17 Summary and Future Research Directions

We have derived in these lecture notes a direct link between the time-reversal technique and the adjoint method for imaging. Using this relationship, we have constructed several iterative time-reversal schemes for solving an inverse problem which arises in ocean acoustic and wireless communication. Each of these schemes can be realized physically as a series of time-reversal experiments, without the use of heavy signal processing or computations. One of the schemes which we have derived (and which we call the 'basic scheme'), is practically equivalent to a technique introduced earlier in [34, 35] using different tools. Therefore, we have given an alternative theoretical derivation of that technique, with a different mathematical interpretation. The other schemes which we have introduced are new in this application. They represent either generalizations of the basic scheme, or alternatives which follow different objectives.

Many questions related to these and similar iterative time-reversal approaches for telecommunication are still open. The experimental implementation has been investigated so far only for one of these techniques in [34, 35], for the situation of underwater sound propagation. A thorough experimental (or, alternatively, numerical) verification of the other schemes is necessary for their practical evaluation. An interesting and practically important problem is the derivation of quantitative estimates for the expected focusing quality of each of these schemes, for example following the ideas of the work performed for a single step in [5]. Certainly, it is expected that these estimates will again strongly depend on the randomness of the medium, on the geometry and distribution of users and base antennas, and on technical constraints as for example partial measurements. Also, different types of noise in the communication system need to be taken into account. The performance in a time-varying environment is another interesting issue of practical importance. All schemes presented here can be adapted in principle to a dynamic environment by re-adjusting the constructed optimal signals periodically. Practical ways of doing so need to be explored.

Acknowledgments

These lecture notes have been written during the stay of the author at L2S-Supélec, France, in the spring of 2004, and are inspired by many discussions with G. Papanicolaou on time reversal and imaging during the Mathematical Geophysics Summer Schools at Stanford University in 1998–2002 and during the CIME program on 'imaging' in Martina Franca in the fall of 2002. The author thanks Dominique Lesselier for making the research stay at Supélec possible. He thanks George Papanicolaou, Hongkai Zhao and Knut Solna for many exciting and useful discussions on time-reversal. He thanks Mathias Fink, Claire Prada, Gabriel Montaldo, Francois Vignon, and the group at Laboratoire Ondes et Acoustique in Paris for pointing out to him their recent work on iterative communication schemes, which stimulated part of the work presented here. The author also thanks Bill Kuperman and the group at Scripps Institution of Oceanography in San Diego for useful discussions. Financial support from the Office of Naval Research (under grant N00014-02-1-0090) and the CNRS (under grant No. 8011618) is gratefully acknowledged.

Appendix A: Proof of Theorem 9.1

For arbitrary $\mathbf{r} \in Z$ and $\mathbf{u} \in U$ we have

$$\langle M\mathbf{u}, \mathbf{r} \rangle_Z = \sum_{k=1}^{K} \int_{[0,T]} \left\langle \int_{\Omega} \gamma_k(\mathbf{x})\mathbf{u}(\mathbf{x},t)d\mathbf{x}, \, \mathbf{r}(t) \right\rangle_N dt$$

$$= \int_{[0,T]} \int_{\Omega} \left\langle \Gamma(\mathbf{x})\mathbf{u}(\mathbf{x},t), \, \Gamma^{-1}(\mathbf{x}) \sum_{k=1}^{K} \gamma_k(\mathbf{x})\mathbf{r}(t) \right\rangle_N d\mathbf{x}dt.$$

Therefore,

$$\langle M\mathbf{u}, \mathbf{r} \rangle_Z = \langle \mathbf{u}, M^*\mathbf{r} \rangle_U$$

with

$$M^*\mathbf{r} = \Gamma^{-1}(\mathbf{x}) \sum_{k=1}^{K} \gamma_k(\mathbf{x})\mathbf{r}(t) = \Gamma^{-1}(\mathbf{x})Q\mathbf{r}.$$

Doing the analogous calculation for $\hat{M}^*$ we get

$$M^* = \Gamma^{-1}Q, \qquad \hat{M}^* = \Gamma^{-1}\hat{Q}. \tag{85}$$

Taking the adjoint of (85) we see that

$$Q^* = M\Gamma, \qquad \hat{Q}^* = \hat{M}\Gamma.$$

holds as well. This proves the theorem.

Appendix B: Proof of Theorem 9.2

We have the following version of *Green's formula*

$$\int_{[0,T]} \int_\Omega \left\langle \Gamma(\mathbf{x})\frac{\partial \mathbf{u}}{\partial t} + \sum_{i=1}^3 D^i \frac{\partial \mathbf{u}}{\partial x_i} + \Phi(\mathbf{x})\mathbf{u}, \mathbf{z} \right\rangle_N dxdt \qquad (86)$$

$$+ \int_{[0,T]} \int_\Omega \langle \Gamma(\mathbf{x})\mathbf{u}(\mathbf{x},t), \mathbf{v}(\mathbf{x},t)\rangle_N \, dxdt + \int_{[0,T]} \int_\Omega \langle \mathbf{q}(\mathbf{x},t), \mathbf{z}(\mathbf{x},t)\rangle_N \, dxdt$$

$$= \int_{[0,T]} \int_\Omega \left\langle \mathbf{u}, -\Gamma(\mathbf{x})\frac{\partial \mathbf{z}}{\partial t} - \sum_{i=1}^3 D^i \frac{\partial \mathbf{z}}{\partial x_i} + \Phi(\mathbf{x})\mathbf{z} \right\rangle_N dxdt$$

$$+ \int_{[0,T]} \int_\Omega \langle \Gamma(\mathbf{x})\mathbf{u}(\mathbf{x},t), \mathbf{v}(\mathbf{x},t)\rangle_N \, dxdt + \int_{[0,T]} \int_\Omega \langle \mathbf{q}(\mathbf{x},t), \mathbf{z}(\mathbf{x},t)\rangle_N \, dxdt$$

$$+ \int_\Omega \langle \Gamma(\mathbf{x})\mathbf{u}(\mathbf{x},T), \mathbf{z}(\mathbf{x},T)\rangle_N \, dx - \int_\Omega \langle \Gamma(\mathbf{x})\mathbf{u}(\mathbf{x},0), \mathbf{z}(\mathbf{x},0)\rangle_N \, dx$$

$$+ \int_{[0,T]} \int_{\partial\Omega} \sum_{i=1}^3 \langle D^i \mathbf{u}, \mathbf{z}\rangle_N \, \nu_i(\mathbf{x}) \, d\sigma dt,$$

where $\mathbf{n}(\mathbf{x}) = (\nu_1(\mathbf{x}), \nu_2(\mathbf{x}), \nu_3(\mathbf{x}))$ is the outward normal at $\partial\Omega$ in the point $\mathbf{x}$. Notice that we have augmented Green's formula in (86) by some terms which appear in identical form on the left hand side and on the right hand side.

We will assume here that the boundary is far away from the sources and receivers and that no energy enters Ω from the outside, such that during the time interval of interest $[0,T]$ all fields along this boundary are identically zero. This is expressed by the boundary conditions given in (4) and (19). Let $\mathbf{u}(\mathbf{x},t)$ be a solution of (1), (2), (4), and $\mathbf{z}(\mathbf{x},t)$ a solution of (17)–(19). Then the first term on the left hand side of (86) and the third term on the right hand side cancel each other because of (1). The second term on the left hand side and the first term on the right hand side cancel each other because of (17). The $(t = T)$-term and the $(t = 0)$ term vanish due to (18) and (2), respectively, and the boundary integral vanishes because of (4) and (19). The remaining terms (i.e. the third term on the left hand side and the second term on the right hand side) can be written as

$$\langle F\mathbf{q}, \mathbf{v}\rangle_U = \langle \mathbf{q}, F^*\mathbf{v}\rangle_U,$$

with $F^*\mathbf{v} = \Gamma^{-1}(\mathbf{x})\mathbf{z}(\mathbf{x},t)$ as defined in (20).

Appendix C: Direct Proof of Theorem 10.1

We prove the theorem by using *Greens formula*:

$$\int_0^T \int_\Omega \left[\rho \frac{\partial \mathbf{v}_f}{\partial t} \mathbf{v}_a + \mathbf{grad} p_f \mathbf{v}_a + \kappa \frac{\partial p_f}{\partial t} p_a + \mathrm{div}\mathbf{v}_f p_a \right] dxdt \qquad (87)$$

$$+ \int_0^T \int_\Omega [\rho \mathbf{v}_f \phi + \kappa p_f \psi] \, dxdt + \int_0^T \int_\Omega [\mathbf{q}_v \mathbf{v}_a + \mathbf{q}_p p_a] \, dxdt$$

$$= \int_0^T \int_\Omega \left[-\mathbf{v}_f \rho \frac{\partial \mathbf{v}_a}{\partial t} - p_f \mathrm{div} \mathbf{v}_a - p_f \kappa \frac{\partial p_a}{\partial t} - \mathbf{v}_f \mathbf{grad} p_a \right] d\mathbf{x} dt$$

$$+ \int_0^T \int_\Omega [\rho \mathbf{v}_f \phi + \kappa p_f \psi] \, d\mathbf{x} dt + \int_0^T \int_\Omega [\mathbf{q}_\mathbf{v} \mathbf{v}_a + \mathbf{q}_p p_a] \, d\mathbf{x} dt$$

$$+ \int_0^T \int_{\partial\Omega} (\mathbf{v}_a \cdot \mathbf{n}) p_f \, d\sigma dt + \int_0^T \int_{\partial\Omega} (\mathbf{v}_f \cdot \mathbf{n}) p_a \, d\sigma dt$$

$$+ \int_\Omega \rho \big[(\mathbf{v}_f \mathbf{v}_a)(\mathbf{x}, T) - (\mathbf{v}_f \mathbf{v}_a)(\mathbf{x}, 0) \big] \, d\mathbf{x} + \int_\Omega \kappa \big[(p_f p_a)(\mathbf{x}, T) - (p_f p_a)(\mathbf{x}, 0) \big] \, d\mathbf{x}.$$

This equation has the form (86). Notice that we have augmented Green's formula in (87), as already shown in (86), by some terms which appear in identical form on the left hand side and on the right hand side.

The first term on the left hand side of (87) and the third term on the right hand side cancel each other due to (21), (22). The second term on the left hand side and the first term on the right hand side cancel each other because of (24), (25). The ($t = T$)-terms and the ($t = 0$)-terms vanish due to (26), (23), respectively, and the boundary terms vanish because of zero boundary conditions. We are left over with the equation

$$\int_0^T \int_\Omega [\rho \mathbf{v}_f \phi + \kappa p_f \psi] \, d\mathbf{x} dt = \int_0^T \int_\Omega [\mathbf{q}_\mathbf{v} \mathbf{v}_a + \mathbf{q}_p p_a] \, d\mathbf{x} dt$$

Defining F^* by (27), this can be written as

$$\left\langle F \begin{pmatrix} \mathbf{q}_\mathbf{v} \\ \mathbf{q}_p \end{pmatrix}, \begin{pmatrix} \phi \\ \psi \end{pmatrix} \right\rangle_U = \left\langle \begin{pmatrix} \mathbf{q}_\mathbf{v} \\ \mathbf{q}_p \end{pmatrix}, F^* \begin{pmatrix} \phi \\ \psi \end{pmatrix} \right\rangle_U.$$

Therefore, F^* is in fact the adjoint of F, and the theorem is proven.

Appendix D: Direct Proof of Theorem 11.1

We prove the theorem by using *Green's formula*:

$$\int_0^T \int_\Omega \left[\epsilon \frac{\partial \mathbf{E}_f}{\partial t} \mathbf{E}_a - \mathbf{curl} \mathbf{H}_f \mathbf{E}_a + \sigma \mathbf{E}_f \mathbf{E}_a + \mu \frac{\partial \mathbf{H}_f}{\partial t} \mathbf{H}_a + \mathbf{curl} \mathbf{E}_f \mathbf{H}_a \right] d\mathbf{x} dt$$

$$+ \int_0^T \int_\Omega [\epsilon \mathbf{E}_f \phi + \mu \mathbf{H}_f \psi] \, d\mathbf{x} dt + \int_0^T \int_\Omega [\mathbf{q}_E \mathbf{E}_a + \mathbf{q}_H \mathbf{H}_a] \, d\mathbf{x} dt$$

$$= \int_0^T \int_\Omega \left[-\mathbf{E}_f \epsilon \frac{\partial \mathbf{E}_a}{\partial t} - \mathbf{H}_f \mathbf{curl} \mathbf{E}_a + \mathbf{E}_f \sigma \mathbf{E}_a - \mathbf{H}_f \mu \frac{\partial \mathbf{H}_a}{\partial t} + \mathbf{E}_f \mathbf{curl} \mathbf{H}_a \right] d\mathbf{x} dt$$

$$+ \int_0^T \int_\Omega [\epsilon \mathbf{E}_f \phi + \mu \mathbf{H}_f \psi] \, d\mathbf{x} dt + \int_0^T \int_\Omega [\mathbf{q}_E \mathbf{E}_a + \mathbf{q}_H \mathbf{H}_a] \, d\mathbf{x} dt$$

$$+ \int_0^T \int_{\partial\Omega} \mathbf{E}_a \times \mathbf{H}_f \cdot \mathbf{n} \, d\sigma dt + \int_0^T \int_{\partial\Omega} \mathbf{E}_f \times \mathbf{H}_a \cdot \mathbf{n} \, d\sigma dt \qquad (88)$$

$$+ \int_\Omega \epsilon \big[(\mathbf{E}_f \mathbf{E}_a)(T) - (\mathbf{E}_f \mathbf{E}_a)(0)\big] \, d\mathbf{x} + \int_\Omega \mu \big[(\mathbf{H}_f \mathbf{H}_a)(T) - (\mathbf{H}_f \mathbf{H}_a)(0)\big] \, d\mathbf{x}$$

This equation has the form (86). Notice that we have augmented Green's formula in (88), as already shown in (86), by some terms which appear in identical form on the left hand side and on the right hand side.

The first term on the left hand side of (88) and the third term on the right hand side cancel each other because of (40) and (41). The second term on the left hand side and the first term on the right hand side cancel each other because of (43), (44). The $(t = 0)$-terms and the $(t = T)$-terms vanish due to (42) and (45). The boundary terms vanish because of zero boundary conditions. We are left over with the equation

$$\int_0^T \int_\Omega \big[\epsilon \mathbf{E}_f \phi + \mu \mathbf{H}_f \psi\big] \, d\mathbf{x} dt = \int_0^T \int_\Omega \big[\mathbf{q}_E \mathbf{E}_a + \mathbf{q}_H \mathbf{H}_a\big] \, d\mathbf{x} dt.$$

Defining F^* by (46), this can be written as

$$\left\langle F \begin{pmatrix} \mathbf{q}_E \\ \mathbf{q}_H \end{pmatrix}, \begin{pmatrix} \phi \\ \psi \end{pmatrix} \right\rangle_U = \left\langle \begin{pmatrix} \mathbf{q}_E \\ \mathbf{q}_H \end{pmatrix}, F^* \begin{pmatrix} \phi \\ \psi \end{pmatrix} \right\rangle_U.$$

Therefore, F^* is in fact the adjoint of F, and the theorem is proven.

References

1. Bal G and Ryzhik L 2003 Time reversal and refocusing in random media *SIAM J. Appl. Math.* **63** 1475–98
2. Bardos C and Fink M 2002 Mathematical foundations of the time reversal mirror *Asymptotic Analysis* **29** 157–182
3. Berryman J G, Borcea L, Papanicolaou G C and Tsogka C 2002 Statistically stable ultrasonic imaging in random media *Journal of the Acoustic Society of America* **112** 1509–1522
4. Bleistein N, Cohen J K, and Cohen J W 2001 *Mathematics of multidimensional seismic imaging, migration, and inversion* (Springer: New York)
5. Blomgren P, Papanicolaou G, and Zhao H 2002 Super-Resolution in Time-Reversal Acoustics, *Journal of the Acoustical Society of America* **111** 238–248
6. Borcea L, Papanicolaou G and Tsogka C 2003 Theory and applications of time reversal and interferometric imaging *Inverse Problems* **19** 5139–5164
7. Cheney M, Isaacson D and Lassas M 2001 Optimal Acoustic Measurements *SIAM J Appl Math*, **61** 1628–1647
8. Claerbout J 1985 *Fundamentals of geophysical data processing: with applications to petroleum prospecting*, (CA: Blackwell Scientific Publications, Palo Alto)
9. Collins M D and Kuperman W A 1991 Focalization: Environmental focusing and source localization *J. Acoust. Soc. Am.* **90** (3) 1410–1422
10. Collins M D and Kuperman W A 1994 Inverse problems in ocean acoustics *Inverse Problems* **10** 1023–1040

11. Dautray R and Lions J L *Mathematical Analysis and Numerical Methods for Science and Technology* Vol. 1–6, Springer.

12. Devaney A 2004 Superresolution processing of multi-static data using time reversal and MUSIC *to appear in J. Acoust. Soc. Am.*

13. Dorn O 1998 A transport-backtransport method for optical tomography *Inverse Problems* **14** 1107–1130

14. Dórn O, Bertete-Aguirre H, Berryman J G and Papanicolaou G C 1999 A nonlinear inversion method for 3D electromagnetic imaging using adjoint fields *Inverse Problems* **15** 1523-1558

15. Dowling D R and Jackson D R 1992 Narrow band performance of phase conjugate arrays in dynamic random media *J. Acoust. Soc. Am.* **91** 3257-3277

16. Dowling D R 1993 Acoustic pulse compression using passive phase-conjugate processing *J. Acoust. Soc. Am.* **95** (3) 1450–1458

17. Draeger C and Fink M 1997 One-Channel Time Reversal of Elastic Waves in a Chaotic 2D-Silicon Cavity *Physical Review Letters* **79** (3) 407–410

18. Edelmann G F, Akal T, Hodgkiss W S, Kim S, Kuperman W A and Chung H C 2002 An initial demonstration of underwater acoustic communication using time reversal *IEEE J. Ocean. Eng.* **27** (3) 602–609

19. Emami S M, Hansen J, Kim A D, Papanicolaou G, Paulraj A J, Cheung D and Prettie C 2004 Predicted time reversal performance in wireless communications using channel measurements *submitted*

20. Fink M 1997 Time reversed acoustics *Phys. Today* **50** 34-40

21. Fink M and Prada C 2001 Acoustic time-reversal mirrors *Inverse Problems* **17** R1–R38

22. Foschini G J 1996 Layered space-time architecture for wireless communications in a fading environment *Bell Labs Technical Journal* **1** (2) 41–59

23. Fouque J P, Garnier J and Nachbin A 2004 Time reversal for dispersive waves in random media *to appear in SIAM J. Appl. Math.*

24. Gesbert D, Shafi M, Shiu D, Smith P J and Naguib A 2003 From Theory to Practice: An Overview of MIMO Space-Time Coded Wireless Systems *IEEE J. Select. Areas Commun.* **21** (3) (tutorial paper) 281–301

25. Haber E, Ascher U and Oldenburg D 2000 On optimization techniques for solving nonlinear inverse problems *Inverse Problems* **16** 1263–1280

26. Hodgkiss W S, Song H C, Kuperman W A, Akal T, Ferla C and Jackson D R 1999 A long range and variable focus phase conjugation experiment in shallow water *J. Acoust. Soc. Am.* **105** 1597–1604

27. Hursky P, Porter M B, Cornuelle B D, Hodgkiss W S, and Kuperman W A 2004 Adjoint modeling for acoustic inversion *J. Acoust. Soc. Am.* **115** (2) 607–619

28. Jackson D R and Dowling D R 1991 Phase conjugation in underwater acoustics *J. Acoust. Soc. Am.* **89** (1) 171–181

29. Kim A D, Blomgren P and Papanicolaou G 2004 Spatial focusing and intersymbol interference in time reversal *submitted*

30. Klibanov M V and Timonov A 2003 On the mathematical treatment of time reversal *Inverse Problems* **19** 1299–1318

31. Kuperman W A, Hodgkiss W S, Song H C, Akal T, Ferla C, and Jackson D R 1998 Phase conjugation in the ocean: Experimental demonstration of an acoustic time reversal mirror *J. Acoust. Soc. Am.* **103** 25-40

32. Kyritsi P, Papanicolaou G, Eggers P and Oprea A 2004 MISO time reversal and delay spread compression for FWA channels at 5GHz *preprint*

33. Marsden J E and Hughes T J R *Mathematical Foundations of Elasticity* (Prentice Hall, 1968)

34. Montaldo G, Lerosey G, Derode A, Tourin A, de Rosny J and Fink M 2004 Telecommunication in a disordered environment with iterative time reversal *Waves Random Media* **14** 287–302
35. Montaldo G, Tanter M and Fink M 2004 Real time inverse filter focusing through iterative time reversal *to appear in J. Acoust. Soc. Am.* **115** (2)
36. Natterer F and Wübbeling F 1995 A propagation-backpropagation method for ultrasound tomography *Inverse Problems* **11** 1225-1232
37. Oestges C, Kim A D, Papanicolaou G and Paulraj A J 2004 Characterization of space-time focusing in time reversal random fields *submitted to IEEE Trans. Antennas Prop.*
38. Papanicolaou G, Ryzhik L and Solna K 2004 Statistical stability in time-reversal *SIAM J. Appl. Math.* **64** 1133–1155
39. Prada C, Manneville S, Spoliansky D and Fink M 1996 Decomposition of the time reversal operator: Detection and selective focusing of two scatterers *J. Acoust. Soc. Am.* **96** (4) 2067–2076
40. Prada C, Kerbrat E, Cassereau D and Fink M 2002 Time reversal techniques in ultrasonic nondestructive testing of scattering media *Inverse Problems* **18** 1761–1773
41. Ryzhik L, Papanicolaou G and Keller J B 1996 Transport Equations for Elastic Waves and Other Waves in Random Media *Wave Motion* **24** 327–370
42. Simon H S, Moustakas A L, Stoytchev M and Safar H 2001 Communication in a disordered world *Physics Today* **54** (9) (online at http://www.physicstoday.org/pt/vol-54/iss-9/p38.html)
43. Snieder R K and Scales J A 1998 Time-reversed imaging as a diagnostic of wave and particle chaos *Phys. Rev. E* **58** (5) 5668–5675
44. Smith K B, Abrantes A A M and Larraza A 2003 Examination of time-reversal acoustics in shallow water and applications to non-coherent underwater communications *J. Acoust. Soc. Am.* **113** (6) 3095–110
45. Solna K 2002 Focusing of time-reversed reflections *Waves Random Media* **12** 1–21
46. Song H, Kuperman W A, Hodgkiss W S, Akal T and Guerrini P 2003 Demonstration of a High-Frequency Acoustic Barrier With a Time-Reversal Mirror *IEEE J. Acoust. Eng.* **28** (2) 246–249
47. Tsogka C and Papanicolaou G C 2002 Time reversal through a solid-liquid interface and super-resolution *Inverse Problems* **18** 1639-1657
48. Vögeler M 2003 Reconstruction of the three-dimensional refractive index in electromagnetic scattering by using a propagation-backpropagation method *Inverse Problems* **19** 739–753
49. Yang T C 2003 Temporal Resolutions of Time-Reversal and Passive-Phase Conjugation for Underwater Acoustic Communications *IEEE J. Ocean. Eng.* **28** (2) 229–245
50. Zhao H 2004 Analysis of the Response Matrix for an Extended Target *SIAM J. Appl. Math.* **64** (3) 725–745

A Brief Review on Point Interactions

Gianfausto Dell'Antonio[1], Rodolfo Figari[2], and Alessandro Teta[3]

[1] Dipartimento di Matematica, Università di Roma "La Sapienza", Rome, Italy and Laboratorio Interdisciplinare SISSA-ISAS, Trieste, Italy
gianfa@sissa.it
[2] Dipartimento di Scienze Fisiche, Universitá di Napoli, Naples, Italy and Istituto Nazionale di Fisica Nucleare, Sezione di Napoli, Naples, Italy
figari@na.infn.it
[3] Dipartimento di Matematica Pura e Applicata, Universitá di L'Aquila, via Vetoio – loc. Coppito, L'Aquila, Italy
teta@univaq.it

Summary. We review properties and applications of point interaction Hamiltonians. This class of operators is first defined following a classical presentation and then generalized to cases in which some dynamical and/or geometrical parameters are varying with time. We recall their relations with smooth short range potentials.

1 Introduction

In this lecture we shall review some basic facts on the so-called point interactions, i.e. perturbations of the free Laplacian in $\mathbb{R}^d$, $d = 1, 2, 3$, supported by a finite set of points (for a comprehensive treatment we refer to the monograph [AGH-KH]).

At a formal level the operator can be written as

$$ \text{``} H = -\Delta + \sum_{j=1}^{n} \alpha_j \delta_{y_j} \text{''} \tag{1} $$

where $\alpha_j \in \mathbb{R}$ and $y_j \in \mathbb{R}^d$ denote respectively the strength and the position of the j-th point interaction.

We shall be concerned with selfadjoint realizations in $L^2(\mathbb{R}^d)$ of the formal expression (1).

Historically such kind of interactions, also called δ-interactions or zero-range interactions or Fermi pseudo-potentials, have been introduced for the Schrödinger equations in the early days of quantum mechanics.

Nevertheless we want to emphasize that they can be equally well introduced in any evolution or stationary problem involving a selfadjoint perturbation of the free Laplacian (or of a regular elliptic operator) in $L^2(\mathbb{R}^d)$ and for hyperbolic evolution equations.

In particular one can also consider applications to the heat and the wave equations.

Physical motivations for the introduction of point interactions were given by Fermi ([F]) in the analysis of scattering of slow neutrons from a target of condensed matter.

In fact, for a sufficiently slow neutron, the wavelength is much larger than the effective range of the nuclear force acting between the neutron and each nucleus of the target and, on the other hand, such force is extremely intense. As a consequence it appears reasonable to modelize the interaction between the neutron and the nucleus by a zero-range potential placed at the position of the nucleus.

As a first approximation one considers each nucleus in a fixed position y_j and this leads to the basic model Hamiltonian (1) and to the corresponding linear evolution for the wave function of the neutrons solving the Schrödinger equation.

The main interest of point interactions consists in the fact that they define simple but non trivial models of short range interactions which are, in fact, explicitly solvable.

More precisely this means that detailed information about the spectrum and the eigenfunctions of (1) are available and then all the physical relevant quantities related to the specific problem analyzed can be explicitly computed.

This fact makes point interactions especially useful in applications where they can be considered as a first step in a more detailed analysis.

More refined evolution models can be obtained considering the strength and/or the position of the point interactions as given functions of time (linear non-autonomous evolution problems) or the strength as a given function of the solution itself (non linear evolution problems).

From a mathematical point of view, the first step is to give a satisfactory definition of the formal operator (1) as a selfadjoint operator in $L^2(\mathbb{R}^d)$ corresponding to the intuitive idea of point interactions.

One can start from the reasonable consideration that any rigorous counterpart of (1) must satisfy $Hu = -\Delta u$ for any $u \in C_0^\infty(\mathbb{R}^d \setminus \{y_1, \ldots, y_n\})$.

This suggests to define the following restriction of the free Laplacian

$$\hat{H} = -\Delta, \quad D(\hat{H}) = C_0^\infty(\mathbb{R}^d \setminus \{y_1, \ldots, y_n\}) \tag{2}$$

It is not hard to see that the operator (2) is symmetric but not selfadjoint in $L^2(\mathbb{R}^d)$; moreover one selfadjoint extension of (2) is trivial, i.e. it corresponds to the free Laplacian $H_0 = -\Delta$, $D(H_0) = H^2(\mathbb{R}^d)$.

This naturally leads to the following definition.

Definition 1. We call Laplacian with point interactions placed at $y_1, \ldots, y_n \in \mathbb{R}^d$ any non trivial selfadjoint extension of (2).

Many different mathematical techniques have been used to construct and classify such class of operators.

A general approach is based on the theory of selfadjoint extensions of symmetric operators, developed by von Neumann and Krein, but also approximation procedures in the resolvent sense, non-standard analysis or the theory of quadratic forms can be used. We shall outline the construction based on the theory of selfadjoint extensions.

The paper is organized as follows.

In Sect. 2 we consider the case of one or many interaction centers in $d = 3$ and $d = 1$. Few details will be given, mainly referring to the particularly simple case of a single point interaction in dimension $d = 3$. (We shall omit to present the $d = 2$ case which, in many respects, is analogous to the three dimensional case.)

In Sect. 3 we consider the connections between point interaction Hamiltonians and Schrödinger operators with smooth potentials.

In Sect. 4 we introduce time dependent and non linear point interactions and we discuss possible applications of such kind of operators.

2 Construction of Point Interactions

Let us denote with $(\cdot, \cdot)$ the inner product in $L^2(\mathbb{R}^d)$. As a consequence of the definition (2), for any function $\psi \in H^2(\mathbb{R}^3)$ and $\phi \in D(\hat{H})$, we have

$$(\psi, -\Delta\phi) = (-\Delta\psi, \phi)$$

and

$$|(-\Delta\psi, \phi)| \le \|\psi\|_{H^2}\|\phi\|_{L^2}$$

implying that all functions in $H^2(\mathbb{R}^3)$ belong to the domain of the operator $\hat{H}^*$ adjoint to $\hat{H}$.

On the other hand for any $z \in \mathbb{C} \setminus \mathbb{R}^+$ we have

$$(G_i^z, \hat{H}\phi) = (zG_i^z, \phi) \qquad \phi \in D(\hat{H}), \qquad i = 1, \ldots, n \qquad (3)$$

where G_i^z denotes the inverse Fourier transform of

$$\tilde{G}_i^z(k) = \frac{e^{ik \cdot y_i}}{k^2 - z}$$

In fact the G_i^z satisfy, in the sense of distributions, $(-\Delta - z)G_i^z = \delta_{y_i}$. Being the δ_{y_i} together with their derivatives the only distributions supported by the discrete set $\{y_1, \ldots, y_n\}$ the G_i^z and their derivatives are the only distributions satisfying (3).

We conclude that the eigenspace $N_z \subset L^2(\mathbb{R}^d)$ relative to the eigenvalue z of the adjoint $\hat{H}^*$ of $\hat{H}$ contains all the G_i^z and their derivatives as long as they belong to $L^2(\mathbb{R}^d)$.

Notice that for $z \in \mathbb{C} \setminus \mathbb{R}^+$:

- In $d = 1$ the G_i^z and their first derivatives belong to $L^2(\mathbb{R})$.
- In $d = 2, 3$ the G_i^z belong to $L^2(\mathbb{R}^d)$ but their derivatives do not.
- In $d > 3$ no function in $L^2(\mathbb{R}^d)$ satisfies (3).

Summarizing the considerations made before we see that, for a finite number of point interactions, the subspaces N_z have dimension which is constant as z varies over the complex plane, outside the positive real axis and it is equal to $2n$ in $d = 1$, to n in $d = 2, 3$ and to 0 for $d > 3$; Moreover the domain of the adjoint operator $\hat{H}^*$ contains $H^2(\mathbb{R}^d)$ and all the subspaces N_z for z in the complex plane outside the positive real axis.

Notice that the difference $G_i^{z_1} - G_i^{z_2}$ for $z_1, z_2 \in \mathbb{C} \setminus \mathbb{R}^+$ belongs to $H^2(\mathbb{R}^d)$. On one side this means that the linear span $\mathcal{D}$ generated by linear combinations of functions $f \in H^2(\mathbb{R}^d)$ and $g \in N_{\hat{z}}$, for some $\hat{z}$, includes all linear combinations of the same kind with $g \in N_z$ for any $z \in \mathbb{C} \setminus \mathbb{R}^+$.

On the other side any function in $\mathcal{D}$ can be alternatively expressed as a linear combination of functions in N_{z_1}, N_{z_2} and $H^2(\mathbb{R}^3 \setminus \{y_1, \ldots, y_n\})$ for any choice of $z_1, z_2 \in \mathbb{C} \setminus \mathbb{R}^+$. In fact the value taken at y_i by the regular part of a linear combination $\alpha_1 G_i^{z_1} + \alpha_2 G_i^{z_2}$, i.e.

$$\lim_{x \to y_i} \left(\alpha_1 G_i^{z_1}(x) + \alpha_2 G_i^{z_2}(x) - \frac{\alpha_1 + \alpha_2}{4\pi|x - y_i|} \right)$$

can assume any complex value as α_1 and α_2 vary.

It is not hard to prove that in fact $\mathcal{D}$ is the entire domain of $\hat{H}^*$. This result is a particular case of a general decomposition formula (often quoted as von Neumann formula) stating that if A is a densely defined symmetric operator in a separable Hilbert space than the domain of its adjoint operator is completely characterized as follows:
for all z such that $\Im z > 0$, any vector f in the domain of A^* has a unique decomposition

$$f = f_0 + f^z + f^{\bar{z}} \tag{4}$$

where f_0 belongs to the domain of the closure of A and f^z (resp. $f^{\bar{z}}$) belongs to the eigenspace of A^* relative to the eigenvalue z (resp. $\bar{z}$). The action of A^* on f is obviously given by

$$A^* f = A f_0 + z f^z + \bar{z} f^{\bar{z}} \tag{5}$$

As we mentioned before decomposition (4) can be easily proved directly in our specific case. The proof in the general case can be found, e.g. in [AG].

An immediate application of the above result to our case is that there are no extensions of $\hat{H}$ different from the Laplacian in dimension $d > 3$.

Formula (5) suggests the strategy one has to adopt in order to construct a selfadjoint extension of $\hat{H}$. Any such extension must correspond to a choice of a subspace of the linear span of functions in N_z and in $N_{\bar{z}}$ on which $\hat{H}^*$ acts as a selfadjoint operator (symmetric as long as the dimension of N_z is finite).

Let us follow the details of this construction in the case of one point interaction in $d = 3$ placed at the point $y \in \mathbb{R}^3$. We recall that for $d = 3$

$$G^z(x) = \frac{\exp i\sqrt{z}|x|}{4\pi|x|}$$

From (4) and (5) we have in this case

$$D(\hat{H}^*) = \{f \in L^2(\mathbb{R}^3)| f = f_0 + \beta G^z(\cdot - y) + \gamma G^{\bar{z}}(\cdot - y)\}$$

where $f_0 \in H^2(\mathbb{R}^3 \setminus \{y\})$ and

$$\hat{H}^* f = -\Delta f_0 + \beta z G^z(\cdot - y) + \gamma \bar{z} G^{\bar{z}}(\cdot - y)$$

A direct computation gives

$$\left([\beta G^z(\cdot - y) + \gamma G^{\bar{z}}(\cdot - y), \hat{H}^*[\beta G^z(\cdot - y) + \gamma G^{\bar{z}}(\cdot - y)]\right) = \qquad (6)$$
$$= (|\beta|^2 z + |\gamma|^2 \bar{z}) \|G^z\|^2 + 2\Re(\bar{\gamma}\beta(G^{\bar{z}}, G^z))$$

showing that $\hat{H}^*$ acts as a symmetric operator on linear combinations of G^z and $G^{\bar{z}}$ if and only if $|\beta| = |\gamma|$. We conclude that there are infinitely many selfadjoint extension of $\hat{H}$ (equivalently selfadjoint restrictions of $\hat{H}^*$) defined in the following way

$$D(H_{\phi,y}) = \{f \in L^2(\mathbb{R}^3)| f = f_0 + \beta G^z(\cdot - y) + \beta e^{i\phi} G^{\bar{z}}(\cdot - y)\}$$

$$H_{\phi,y} f = -\Delta f_0 + \beta z G^z + \beta e^{i\phi} \bar{z} G^{\bar{z}} \qquad \qquad (7)$$

An alternative description of the family of selfadjoint extensions $H_{\phi,y}$ is obtained in the following way: take λ positive and define $G_\lambda = G^{z=-\lambda}$. Notice that, around $x = y$

$$G^z(x - y) + e^{i\phi} G^{\bar{z}}(x - y) - (1 + e^{i\phi})G_\lambda(x - y) =$$
$$= \frac{i\sqrt{z}}{4\pi} + e^{i\phi} \frac{i\sqrt{\bar{z}}}{4\pi} + \frac{\sqrt{\lambda}}{4\pi}(1 + e^{i\phi}) + f_0$$
$$= (1 + e^{i\phi})(\alpha + \frac{\sqrt{\lambda}}{4\pi}) + f_0 \qquad \qquad (8)$$

where f_0 is a regular function taking value zero in $x = y$ and

$$\alpha = \frac{\Re\sqrt{z}}{4\pi} \tan\frac{\phi}{2} - \frac{\Im\sqrt{z}}{4\pi}$$

Formula (8) allows to characterize functions in the domains of the different selfadjoint extensions as a relation connecting the behavior of the functions at the singularity (which is $\beta(1 + e^{i\phi})/4\pi|x - y|$ in our case) and the value taken at the same point by their regular part. More precisely (8) implies that for any $\alpha \in \mathbb{R}$ there is a selfadjoint extension of $\hat{H}$ defined in the following way

$$D(H_{\alpha,y}) =$$
$$= \left\{ f \in L^2(\mathbb{R}^3) | f = \Phi_\lambda + qG_\lambda(\cdot - y), \quad \Phi_\lambda \in H^2(\mathbb{R}^3), \quad q = \frac{\Phi_\lambda(0)}{\alpha + \sqrt{\lambda}/4\pi} \right\} \quad (9)$$

Taking into account that

$$-\Delta \left(G^z(x-y) + e^{i\phi}G^{\bar{z}}(x-y) - (1+e^{i\phi})G_\lambda(x-y) \right) =$$
$$= zG^z(x-y) + e^{i\phi}\bar{z}G^{\bar{z}}(x-y) + (1+e^{i\phi})\lambda G_\lambda(x-y)$$

the action of $H_{\alpha,y}$ on $D(H_{\alpha,y})$ is easily found to be

$$(H_{\alpha,y} + \lambda)f = (-\Delta + \lambda)\Phi_\lambda \quad (10)$$

Characterization (9) of the domain implies that the action of $(H_{\alpha,y} + \lambda)^{-1}$ does not differ from the action of $(-\Delta + \lambda)^{-1}$ on functions of $H^2(\mathbb{R}^3 \setminus \{y\})$ (being $q = 0$ in this case).

On the other hand

$$((H_{\alpha,y} + \lambda)^{-1} - (-\Delta + \lambda)^{-1})G_\lambda, (\hat{H} + \lambda)g) = 0$$

for any g in the set $C_0^\infty(\mathbb{R}^3 \setminus \{y\})$, which is dense in $L^2(\mathbb{R}^3)$. As a consequence, the function $((H_{\alpha,y} + \lambda)^{-1} - (-\Delta + \lambda)^{-1})G_\lambda$ belongs to $N_{-\lambda}$, which in turn means that it is proportional to G_λ itself.

The action of the two resolvents then differ only on the one dimensional subspace generated by G_λ. Being $D(H_{\alpha,y})$ in (9) the range of $(H_{\alpha,y} + \lambda)^{-1}$ one finally obtains

$$[(H_{\alpha,y} + \lambda)^{-1}g](x) = (G_\lambda g)(x) + \frac{1}{\alpha + \sqrt{\lambda}/4\pi}G_\lambda(x-y)(G_\lambda g)(y) \quad (11)$$

Notice that the free Laplacian is obtained formally in the limit $\alpha \to \infty$ showing that one cannot consider α as a coupling constant. Its physical meaning becomes clear studying scattering theory for point interaction Hamiltonians. In this setting one proves that α is the inverse of the scattering length associated to the operator $H_{\alpha,y}$.

From the explicit expression of the resolvent (11) one can easily deduce the spectral properties of $H_{\alpha,y}$:

- The spectrum of $H_{\alpha,y}$ is

$$\sigma(H_{\alpha,y}) = [0, \infty) \quad \alpha \geq 0$$

$$\sigma(H_{\alpha,y}) = \{-16\pi^2\alpha^2\} \cup [0, \infty) \quad \alpha < 0$$

where the continuous part $[0, \infty)$ is purely absolutely continuous.
- For $\alpha < 0$ the only eigenvalue is simple and the corresponding normalized eigenfunction is

$$\psi_\alpha(x) = \sqrt{-2\alpha}\frac{\exp(4\pi\alpha|x-y|)}{|x-y|}.$$

– For any $\alpha \in \mathbb{R}$, corresponding to each positive energy E in the continuous spectrum there are infinitely many generalized eigenfunction

$$\psi_\alpha(k, x) = \frac{1}{(2\pi)^{3/2}} \left(\exp(ik \cdot x) - \frac{\exp iky}{\alpha - i|k|/(4\pi)} \frac{\exp(i|k||x - y|)}{|x - y|} \right) \quad (12)$$

with $|k|^2 = E$.

The explicit form of the spectral decomposition of $H_{\alpha,y}$, in terms of the eigenfunction (in the case $\alpha < 0$) and of the generalized eigenfunctions, allows to write the solution of the Schrödinger equation

$$i\frac{\partial \psi_t}{\partial t} = H_{\alpha,y}\psi_t, \quad (13)$$

corresponding to any initial state in $L^2(\mathbb{R}^3)$, as an integral over the spectral measure. In fact the integral kernel, in configuration space, for the propagator of $H_{\alpha,y}$ can be explicitly computed ([ST]). The same procedure allows to write the kernel of the semigroup generated by $H_{\alpha,y}$ ([ABD]).

An implicit characterization of the propagator will be useful in the next section in order to define time dependent point interactions. A formal inverse Laplace transform of (11) suggests for the solution of the Schrödinger equation (13) the following formula showing a free-propagation contribution and a term representing spherical waves generated at the interaction center

$$\psi_t(x) = (U(t)\psi_0)(x) + i \int_0^t ds\, U(t - s, |x - y|)\, q(s) \quad (14)$$

where $U(t)$ is the propagator of the free unitary group defined by the kernel

$$U(t; x - x') = e^{i\Delta t}(x - x') = \frac{e^{i\frac{|x - x'|^2}{4t}}}{(4\pi i t)^{3/2}}$$

It is easily checked that (14) is the solution of the Schrödinger equation (13) corresponding to an initial condition $\psi_0 \in D(H_{\alpha,y})$ if the function $q(t)$ satisfies the Volterra integral equation

$$q(t) + 4\sqrt{i\pi}\alpha \int_0^t ds\, \frac{q(s)}{\sqrt{t - s}} = 4\sqrt{i\pi} \int_0^t ds\, \frac{(U(s)\psi_0)(y)}{\sqrt{t - s}} \quad (15)$$

More precisely one can prove that if $q(t)$ is the unique solution of (15) then (14) defines a function ψ_t which for any $t \geq 0$ belongs to $D(H_{\alpha,y})$ ($q(t)$ being the coefficient of the singular part of ψ_t) and satisfies (13).

We want to conclude the list of properties of the family of Hamiltonians $H_{\alpha,y}$ recalling the expression of the associated quadratic forms.

Quadratic forms are often taken as an alternative way to define point interactions, a way often preferred to the one used above because of their immediate connection with quantities of physical interest. On $D(H_{\alpha,y})$ the quadratic form uniquely associated to the operator $H_{\alpha,y}$ is

$$F_{\alpha,y}(u) =$$

$$= \int_{\mathbb{R}^3} dx \left[|\nabla(u - qG_\lambda(\cdot - y))|^2 + \lambda|(u - qG_\lambda(\cdot - y))|^2 - \lambda|u|^2 \right] + \left(\alpha + \frac{\sqrt{\lambda}}{4\pi} \right) |q|^2 \tag{16}$$

It is bounded below and closable and defines therefore a Dirichlet form. The domain of its closure is

$$D(F_{\alpha,y}) = \{ u \in L^2(\mathbb{R}^3) \mid u = \phi_\lambda + qG_\lambda(\cdot - y), \ \phi_\lambda \in H^1(\mathbb{R}^3), \ q \in \mathbb{C} \} \tag{17}$$

Notice that the form domain is significantly larger than $H^1(\mathbb{R}^3)$. In particular the form $F_{\alpha,y}$ is not a small perturbation of the one defined by the Laplacian.

We shall now go back to the n centers case in $\mathbb{R}^3$ and summarize the main results about the selfadjoint extensions of the operator defined in (2).

As it was mentioned at the end of Sect. 1 the dimension of the eigenspaces N_z relative to the eigenvalue z of the operator adjoint to $\hat{H}$ is n for any $z \in \mathbb{C} \setminus \mathbb{R}^+$ in the case of n centers. Consider any unitary operator V from N_z to $N_{\bar{z}}$ and denote with N_z^V the subspace of the linear span of functions in N_z and in $N_{\bar{z}}$ generated by the linear combinations of the type $f_z + V f_z$ with $f_z \in N_z$. A computation identical to (7) shows that $\hat{H}^*$ acts as a symmetric operator only if its action is restricted to any N_z^V.

As a consequence in this case selfadjoint extensions of $\hat{H}$ are uniquely associated with the n^2 dimensional family of (complex) matrices unitarily connecting N_z and $N_{\bar{z}}$.

In the literature much attention was given to a particular n dimensional subfamily of selfadjoint extensions called local. In fact, in analogy with the one center case, functions in the domain of the n dimensional subfamily of selfadjoint extensions, corresponding to diagonal unitary matrices V, can be alternatively characterized by a specific behavior around each interaction center. We briefly recall the properties of the operators in this subfamily (for an introduction to non local point interactions see [DG]).

For any $\underline{\alpha} = \{\alpha_1, \ldots, \alpha_n\}$ with $\alpha_i \in \mathbb{R}$, $i = 1, \ldots, n$ and $\underline{y} = \{y_1, \ldots, y_n\}$, $y_i \in \mathbb{R}^3$, $i = 1, \ldots, n$ the operator $H_{\underline{\alpha},\underline{y}}$ defined by

$$D(H_{\underline{\alpha},\underline{y}}) = \left\{ u \in L^2(\mathbb{R}^3) \mid u = \phi_\lambda + \sum_{k=1}^n q_k G_\lambda(\cdot - y_k) \right.$$

$$\phi_\lambda \in H^2(\mathbb{R}^3), \quad \phi_\lambda(y_j) = \sum_{k=1}^n [\Gamma_{\underline{\alpha},\underline{y}}(\lambda)]_{jk} q_k, \ j = 1, ..., n \right\} \tag{18}$$

$$(H_{\underline{\alpha},\underline{y}} + \lambda) u = (-\Delta + \lambda)\phi_\lambda \tag{19}$$

where

$$[\Gamma_{\underline{\alpha},\underline{y}}(\lambda)]_{jk} = \left(\alpha_j + \frac{\sqrt{\lambda}}{4\pi} \right) \delta_{jk} - G_\lambda(y_j - y_k)(1 - \delta_{jk}) \tag{20}$$

is the selfadjoint extension of $\hat{H}$ referred to as the (local) point interaction Hamiltonian with n centers located in $\underline{y}$. In fact for any smooth function $u \in D(H_{\underline{\alpha},\underline{y}})$ vanishing at $y_1, ..., y_n$ one has $q = 0$ and then, from (19), $H_{\alpha,y}u = -\Delta u$.

At each point y_j the elements of the domain satisfy a boundary condition expressed by the last equality in (18). If we define $r_j = |x - y_j|$ it is easy to see that the boundary condition can be equivalently written

$$\lim_{r_j \to 0} \left[\frac{\partial(r_j u)}{\partial r_j} - 4\pi\alpha_j(r_j u) \right] = 0, \quad j = 1, ..., n \tag{21}$$

This explains the term "local" by which this class of extensions is known; indeed due to the special form of the matrix (20) the generalized boundary conditions (21) is placed at each point separately. Any other choice for the symmetric invertible matrix $\Gamma_{\underline{\alpha},\underline{y}}(\lambda)$ leads to an extension which is still local in the terminology commonly employed for differential operators (i.e. operators which do not change the support). The extensions obtained by (20) should be perhaps termed "strongly local" or "local" with respect to the boundary conditions.

In order to find the explicit expression for the resolvent it is sufficient to observe that, for any given $f \in L^2(\mathbb{R}^3)$, the solution u of the equation $(H_{\underline{\alpha},\underline{y}} + \lambda)u = f$ must be written in the form $u = G_\lambda f + qG_\lambda(\cdot - y)$ (see (19)), where the charges q are determined imposing the boundary conditions in (18) (or equivalently (21)). The result of the easy computation is

$$(H_{\underline{\alpha},\underline{y}} + \lambda)^{-1} = G_\lambda + \sum_{j,k=1}^{n} [\Gamma_{\underline{\alpha},\underline{y}}(\lambda)]_{jk}^{-1} G_\lambda(\cdot - y_j)G_\lambda(\cdot - y_k) \tag{22}$$

where $\lambda > 0$ is chosen sufficiently large so that the matrix (20) is invertible.

From the analysis of (22) one can derive that the continuous spectrum of $H_{\underline{\alpha},\underline{y}}$ is purely absolutely continuous and coincides with the positive real axis. The point spectrum consists of (at most) n non positive eigenvalues given by the possible solutions $E \geq 0$ of the equation $det[\Gamma_{\underline{\alpha},\underline{y}}(-E)] = 0$.

Proper and generalized eigenfunctions can be explicitly computed.

Notice that the main tool used in the construction of the selfadjoint extensions $H_{\underline{\alpha},\underline{y}}$ was the specific behavior of the Green's function of the Laplacian at the singularity. Therefore the approach we described above can be adapted to treat stationary or evolution problems related to perturbations supported by points of any elliptic operator of the general form

$$L^V = \sum_{k,j=1}^{d} (i\frac{\partial}{\partial x_k} + A_k)a_{kj}(\underline{x})(i\frac{\partial}{\partial x_j} + A_j) + V(\underline{x})$$

under suitable assumptions on the coefficients and on the potential V.

According to the notation introduced above $L_{\underline{\alpha},\underline{y}}^V$ would denote the perturbation of L^V supported by the points y_i with strength α_i. The usefulness of the definition in the general case is linked with the possibility of finding an explicit expression for the Green's function of the operator L^V (as in the case of the harmonic oscillator).

It is easy to check an additivity property with respect to the potential while on the other hand, in dimension 2 and 3, point interactions are not additive in the sense that

$$\left(L^V_{\underline{\alpha},\underline{y}}\right)_{\underline{\beta},\underline{z}} \neq L^V_{\underline{\alpha}\cup\underline{\beta},\underline{y}\cup\underline{z}}$$

As we mentioned in the introduction to this section the family of selfadjoint extensions of $\hat{H}$ has a richer structure in $d = 1$, due to the fact that the subspaces N_z contain also the first derivatives of the Green's functions of the Laplacian centered in the interaction centers. Referring for simplicity to the case of a single point interaction the dimension of $N_z z \in \mathbb{C}\backslash\mathbb{R}^+$ is equal to two. As a consequence there is a four (real) parameter family of selfadjoint extensions of $\hat{H}$.

Without going into details of the construction (see [AGH-KH] for a complete treatment) we recall here the properties of two particular one parameter subfamilies of selfadjoint extensions referred to as δ and δ' interactions, for reasons that will become clear from their definitions.

The one center (placed in $y \in \mathbb{R}$) delta interaction Hamiltonian $H_{\alpha,y}$ is defined as follows

$$D(H_{\alpha,y}) = \left\{ \phi \in H^1(\mathbb{R}) \cap H^2(\mathbb{R}\setminus\{y\}) \mid \frac{d\phi}{dx}(y^+) - \frac{d\phi}{dx}(y^-) = \alpha\phi(y), \ \alpha \in \mathbb{R} \right\} \tag{23}$$

and $H_{\alpha,y}$ acts on a function in its domain as $-d^2/dx^2$ in each point $x \neq y$.

Before discussing the properties of the operators $H_{\alpha,y}$, in order to stress their relation with the one dimensional delta function, we introduce the associated quadratic forms

$$D(F_{\alpha,y}) = H^1(\mathbb{R}) \tag{24}$$

$$F_{\alpha,y}(u) = \int_{\mathbb{R}} dx|\nabla u|^2 + \alpha|u(y)|^2 \tag{25}$$

The form domain is then the same of the form domain of the Laplacian and in fact standard Sobolev inequalities show that $F_{\alpha,y}$ is a small perturbation of the form associated to the Laplacian (see ([T]) for details). In terms of quadratic forms the generalization to n centers placed in $\{y_1,\ldots,y_n\} \equiv \underline{y}$ of strength $\{\alpha_1,\ldots,\alpha_n\} \equiv \underline{\alpha}$ is immediate

$$D(F_{\underline{\alpha},\underline{y}}) = H^1(\mathbb{R}) \tag{26}$$

$$F_{\underline{\alpha},\underline{y}}(u) = \int_{\mathbb{R}} dx|\nabla u|^2 + \sum_{j=1}^{n} \alpha_j|u(y_j)|^2 \tag{27}$$

Formula (27) shows that, differently from the higher dimensional cases, in $d=1$ point interactions are genuinely additive and that the dynamical parameters α_i play the role of coupling constants.

The explicit form of the integral kernel of the resolvent can be computed; in the single center case one has

$$(H_{\alpha,y} + \lambda)^{-1}(x,x') = \frac{e^{-\sqrt{\lambda}|x-x'|}}{2\sqrt{\lambda}} - \frac{2\alpha\sqrt{\lambda}}{\alpha+2\sqrt{\lambda}} \frac{e^{-\sqrt{\lambda}|x-y|}}{2\sqrt{\lambda}} \frac{e^{-\sqrt{\lambda}|x'-y|}}{2\sqrt{\lambda}}$$

Analyzing the singularities of the resolvent one can easily see that the spectrum $\sigma(H_{\alpha,y})$ of $H_{\alpha,y}$ has the properties

$$\sigma(H_{\alpha,y}) = [0,+\infty), \qquad \alpha \geq 0$$

$$\sigma(H_{\alpha,y} = \left\{-\frac{\alpha^2}{4}\right\} \cup [0,+\infty), \qquad \alpha < 0$$

$[0,\infty)$ being purely absolutely continuous.

In the case $\alpha < 0$ the only eigenvalue $-\frac{\alpha^2}{4}$ is non degenerate and the corresponding normalized eigenfunction is

$$\Psi_\alpha(x) = \sqrt{-\frac{\alpha}{2}} \; e^{\frac{\alpha}{2}|x-y|}$$

We refer to [AGH-KH] for the explicit form of the generalized eigenfunctions. Their knowledge allows to compute any bounded function of the operator $H_{\alpha,y}$; in particular its propagator can be written in closed form ([S]). We want to remind the reader that an array of delta interactions in one dimension were used by Kronig and Penney in 1931 ([KP]) to construct a model of the dynamics of an electron inside a crystal. That model remains, after many years, one of the few completely solvable periodic quantum interaction.

In $d = 1$ the implicit equation for the solution of the Schrödinger equation, corresponding to (14) in $d = 3$, can be easily obtained from the formula of the resolvent operator

$$\psi_t(x) = (U_0(t)\psi_0)(x) - i\alpha \int_0^t ds \, U_0(t-s, |x-y|) \, \psi_s(y) \qquad (28)$$

where U_0 is the free propagator in dimension one.

From (28) one sees that the solution is completely determined by the free evolution of ψ_0 and by $\psi_t(y)$, which in turn, as a function of time, must satisfy the integral equation

$$\psi_t(y) = (U_0(t)\psi_0)(y) - i\alpha \int_0^t ds \, U_0(t-s,0) \, \psi_s(y) \qquad (29)$$

It is easy to check that in fact (28) and (29) define a unitary flow in $L^2(\mathbb{R})$ whose generator is $H_{\alpha,y}$ and therefore they may be used as a definition of $H_{\alpha,y}$.

We want briefly to mention the other subfamily of selfadjoint extensions of $\hat{H}$ in $d = 1$ known as δ' interactions.

For each $\beta \in \mathbb{R}$ and $y \in \mathbb{R}$ the operator $H'_{\beta,y}$ can be defined via the action on a function $f \in L^2(\mathbb{R})$ of its resolvent in the following way

$$(H'_{\beta,y} + \lambda)^{-1})f = (-\Delta + \lambda)^{-1}f + \frac{2\beta\lambda}{2 + \beta\sqrt{\lambda}}G'_\lambda(\cdot - y)(G'_\lambda f)(y) \qquad (30)$$

where G'_λ is the derivative of the Green's function for $z = -\lambda$, $\lambda > 0$

$$G'_\lambda = -\frac{1}{2}e^{-\sqrt{\lambda}|x-y|} \quad x > y$$

$$= \frac{1}{2}e^{-\sqrt{\lambda}|x-y|} \quad x < y$$

and $\lambda \neq (2/\beta)^2$ if $\beta < 0$.

The spectrum of $H'_{\beta,y}$ is easily found to be

$$\sigma(H'_{\beta,y}) = [0, +\infty), \qquad \beta \geq 0$$

$$\sigma(H'_{\beta,y}) = \left\{-\frac{4}{\beta^2}\right\} \cup [0, +\infty), \qquad \beta < 0$$

For further details and properties on this kind of operators see [AGH-KH].

3 Connection with Smooth Interactions

It is natural to think of a point interaction as a limit of rescaled short range potentials. On the other hand it is conceivable that any Hamiltonian with a smooth potential can be approximated by an array of a large number of point interactions of suitable strengths.

Let us discuss the former aspect first.

In dimension one this intuitive picture can be made rigorous under very general conditions. For any $\epsilon > 0$, consider the following approximating operators

$$H_\epsilon = -\Delta + \frac{1}{\epsilon}\sum_{j=1}^{n} V_j(\epsilon^{-1}(x - y_j)) \qquad (31)$$

where the potentials V_j can be chosen in $L^1(\mathbb{R})$. Then for any $\lambda > 0$ sufficiently large one has

$$\lim_{\epsilon \to 0} \|(H_\epsilon + \lambda)^{-1} - (H_{\alpha,y} + \lambda)^{-1}\| = 0 \qquad (32)$$

with $\alpha_j = \int_\mathbb{R} dx V_j(x)$ (see [AGH-KH]).

In dimension three the situation is more delicate and the class of approximating potentials must be properly chosen.

In particular one has to restrict to potentials V such that $-\Delta + V$ has a so-called zero-energy resonance, i.e. there exists a solution $\psi \in L^2_{loc}(\mathbb{R}^3)$, with $\nabla\psi \in L^2(\mathbb{R}^3)$, of the equation $(-\Delta + V)\psi = 0$. We call such a solution a zero-energy resonance function.

To simplify the notation, let us fix $n = 1$ and consider $V \in L^2(\mathbb{R}^3)$, with $(1 + | \cdot |)V \in L^1(\mathbb{R}^3)$; moreover assume that there is a unique $\xi \in L^2(\mathbb{R}^3)$ such that

$$uG_0 v\xi = -\xi, \qquad (v, \xi) \neq 0 \tag{33}$$

where $u(x) = |V(x)|^{1/2} sgn[V(x)]$, $v(x) = |V(x)|^{1/2}$. Condition (33) implies that $\psi(x) = (G_0 v\xi)(x)$ is a zero-energy resonance function.

Now consider the following sequence of approximating operators

$$H_\epsilon = -\Delta + \frac{1 + \epsilon\mu}{\epsilon^2} V(\epsilon^{-1}(x - y)), \quad \mu \in \mathbb{R} \tag{34}$$

Under the above assumptions and for any $\lambda > 0$ sufficiently large one can prove

$$\lim_{\epsilon \to 0} \|(H_\epsilon + \lambda)^{-1} - (H_{\alpha,y} + \lambda)^{-1}\| = 0 \tag{35}$$

where $\alpha = -\mu|(v, \xi)|^{-2}$ (see [AGH-KH]).

It is clear from the rescaling of the potential in (34) that a point interaction in dimension three cannot be considered as a true δ-function. Notice that for $\mu = 0$ one gets $\alpha = 0$, which does not correspond to the free Laplacian (see (22)).

It should be emphasized that if the resonance condition (33) is not fulfilled then in the limit $\epsilon \to 0$ one obtains a trivial result, i.e. the free Laplacian.

The approximation of a point interaction in dimension three can be made more transparent if one considers the connection with the following boundary value problem

$$(-\Delta + \lambda)u_\epsilon = f \qquad \text{in } \mathbb{R}^3 \setminus S_\epsilon \tag{36}$$

$$\frac{\partial u_\epsilon}{\partial n_+} - \frac{\partial u_\epsilon}{\partial n_-} = \gamma_\epsilon u_\epsilon \qquad \text{on } S_\epsilon \tag{37}$$

where $\lambda > 0$, $f \in L^2(\mathbb{R}^3)$, $\gamma_\epsilon \in \mathbb{R}$ and S_ϵ is the sphere of radius ϵ and center y.

The solution of (36), (37) is nothing but $(H_{\gamma_\epsilon,S_\epsilon} + \lambda)^{-1}f$, where $H_{\gamma_\epsilon,S_\epsilon}$ denotes the Laplacian in $\mathbb{R}^3$ with a δ-shell interaction of strength γ_ϵ supported by the sphere S_ϵ.

For $\epsilon \to 0$ the sphere shrinks to the point y; assuming

$$\gamma_\epsilon = -\frac{1}{\epsilon} + 4\pi\alpha \tag{38}$$

and $\lambda > 0$ is sufficiently large one can show that ([FT])

$$\lim_{\epsilon \to 0} \|(H_{\gamma_\epsilon,S_\epsilon} + \lambda)^{-1} - (H_{\alpha,y} + \lambda)^{-1}\| = 0 \tag{39}$$

Let us now go back to the second problem mentioned at the beginning of this section concerning a procedure allowing to approximate (in a way which should be suitable for applications) a Schrödinger operator with smooth potential via point interaction Hamiltonians.

The answer is obviously positive in dimension $d = 1$ for the particular self-adjoint Hamiltonians we referred to as delta interaction Hamiltonians. In that case point interactions are additive and, at least in the sense of quadratic forms, they correspond to a potential which is a sum of delta functions. It is easy to prove that any approximation of a potential function through linear combinations of delta measures turns out to be an efficient approximation scheme for the corresponding operators ([BFT]).

As we mentioned the situation is fairly different in dimension $d = 2$ and $d = 3$. In these cases point interaction Hamiltonians for n centers are in no sense sum of an unperturbed operator and a potential, as we remarked at the end of the last section. Moreover point interactions in $d = 2$ and $d = 3$ are in no sense connected to delta measures as we discussed in the first part of this section.

Nevertheless, with some peculiar differences with respect to the one dimensional case, it is possible to work out an approximation procedure that we outline here restricting again our attention to the case $d = 3$. Let us denote with $Y^N = \{y_1^{(N)}, \ldots, y_N^{(N)}\}$ a configuration of N points in $\mathbb{R}^3$ and consider a (product) probability measure $[\rho(y)dy]^{\otimes N}$ on such configuration space. For simplicity let us assume that the density is continuous. Let $\alpha(y)$ be a real function on $\mathbb{R}^3$ continuous and bounded away from zero outside a set of Lebesgue measure zero and define $\alpha^N \equiv \{\alpha(y_1^{(N)}) \ldots, \alpha(y_N^{(N)})\}$. The following theorem was proved in [FHT]

On a set of configurations of measure increasing to 1 as N tends to infinity, for any λ positive, large enough

$$s - \lim_{N \to \infty} \left(H_{N\alpha^N, Y^N} + \lambda) \right)^{-1} = \left(-\Delta - \frac{\rho}{\alpha} + \lambda \right)^{-1} \quad (40)$$

We want to mention an immediate consequence of the result stated above. Notice first that any smooth integrable potential V can be written in the form ρ/α taking

$$\rho(y) = \frac{|V(y)|}{\int_{\mathbb{R}^3} |V(y)| dy} \qquad \alpha = \frac{(sgnV)(y)}{\int_{\mathbb{R}^3} |V(y)| dy}$$

We conclude that any Schrödinger operator with a smooth integrable potential can be approximated by point interaction Hamiltonians.

As a final remark we want to stress that (40) shows that positive potentials are obtained as limits of negative alphas and vice versa. This is a consequence of the fact that even if all the α_i's are positive the corresponding n centers point interaction Hamiltonian can have negative eigenvalues if some, or all, the points are very close to each other. This depends in turn on the fact that the non diagonal terms in the matrix (20) can dominate the diagonal terms if there are couples of very close points.

On the other hand if all the α_i's are negative the rescaled Hamiltonian shows N negative eigenvalues all tending to infinity for large N. In this second case there is no uniform bound from below for the eiganvalues of the operators in the approximating sequence and in particular no convergence of the corresponding semigroups can take place.

4 Time Dependent Point Interactions

In this last section we shall mention some more recent results obtained for time dependent (linear or non linear) point interactions.

In order to avoid unnecessary technical difficulties we are going to outline only the case of a single interaction center in $d = 1$ and $d = 3$.

We start with the $d = 1$ case noticing that the procedure of rephrasing the Schrödinger evolution problem via formulae (28), (29) is directly generalizable to the case of y and α varying with time. In fact, under suitable conditions on the regularity of $y(t), \alpha(t)$ and on the initial condition it is possible to prove that for each $s \in R$ the map $\psi_s \to \psi_t$ induced by

$$\psi_t(x) = (U_0(t-s)\psi_s)(x) - i \int_s^t d\tau \, \alpha(\tau) \, U_0 \left(t - \tau, |x - y(\tau)| \right) \psi_\tau(y(\tau)) \quad (41)$$

where $\psi_t(y(t))$ satisfies

$$\psi_t(y(t)) = (U_0(t-s)\psi_s)(y(t)) - i \int_s^t d\tau \, \alpha(\tau) \, U_0 \left(t - \tau, |y(t) - y(\tau)| \right) \psi_\tau(y(\tau))$$

$$(42)$$

defines a group of unitary transformations $V(s,t)$, with $V(s,s) = 1$, which solves the Schrödinger equation

$$i \frac{\partial V(s,t)\psi_s}{\partial t} = H_{\alpha(t),y(t)} V(s,t)\psi_s \quad (43)$$

Notice that in dimension one the form domain of the Hamiltonians $H_{\alpha(t),y(t)}$ does not depend on time and coincide with $H^1(\mathbb{R})$. This permits to apply the general abstract theory about the solution of non autonomous evolution problems (see, e.g. [Si]).

In $d = 3$ the problem consists in generalizing the representation formulae (14), (15) to the case in which α or the position of the interaction center are varying in time.

Notice that the form domain in (16) changes if the position of the point y varies. As a consequence the Schrödinger problem corresponding to a moving point interaction is more complicate than in dimension one. The problem in the case where the strength parameter α changes in time (in which case the operator domain is varying but not the form domain) was solved in [SY]. In the following theorem the results in the general case when strength and position change in time are summarized.

If $y(t)$ is regular curve in $\mathbb{R}^3$, $\alpha(t)$ a smooth function in $\mathbb{R}$ and $f \in C_0^\infty(\mathbb{R}^3 \setminus \{y(s)\})$ then there exists a unique $\psi_s(t) \in D(F_{\alpha(t),y(t)})$, $t \in \mathbb{R}$, such that $\partial\psi_s(t)/\partial t$ is in the dual (with respect to $L^2(\mathbb{R}^3)$) of $D(F_{\alpha(t),y(t)})$ and

$$i < v(t), \frac{\partial\psi_s(t)}{\partial t} > = B_{\alpha(t),y(t)}(v(t), \psi_s(t)) \quad \forall v(t) \in D(F_{\alpha(t),y(t)})$$

$$\psi_s(s) = f$$

$$(44)$$

where $B_{\alpha(t),y(t)}$ is the bilinear form corresponding to the quadratic form $F_{\alpha(t),y(t)}$.

Moreover $\psi_s(t)$ has the following representation

$$\psi_s(t) = U(t-s)f + i \int_s^t d\tau\, U(t-\tau; \cdot - y(\tau))q(\tau) \tag{45}$$

where the charges $q(t)$ satisfy the Volterra integral equation

$$q(t) + \frac{4\sqrt{\pi}}{\sqrt{-i}} \int_s^t d\tau\, \alpha(\tau) \frac{q(\tau)}{\sqrt{t-\tau}} + \int_s^t d\tau q(\tau)C(t,\tau) = \frac{4\sqrt{\pi}}{\sqrt{-i}} \int_s^t d\tau \frac{(U_0(\tau-s)f)(y(\tau))}{\sqrt{t-\tau}} \tag{46}$$

where

$$C(t,\tau) = -\frac{1}{\pi} \int_0^1 dz \frac{1}{\sqrt{(1-z)z}} \left(A(\tau + (t-\tau)z, \tau) + \dot{B}(\tau + (t-\tau)z, \tau) \right)$$
$$+ \frac{B(\tau + (t-\tau)z, \tau) - 1}{2(t-\tau)z} \Bigg)$$

$$A(t,\tau) = i \frac{(y(t) - y(\tau)) \cdot \dot{y}(\tau)}{2(t-\tau)} \frac{1}{w^3(t,\tau)} \int_0^{w(t,\tau)} dz\, z^2 e^{iz^2}$$

$$B(t,\tau) = \frac{1}{w(t,\tau)} \int_0^{w(t,\tau)} dz\, e^{iz^2}$$

$$w(t,\tau) = \frac{|y(t) - y(\tau)|}{2\sqrt{t-\tau}}$$

and $\dot{B}(t,\tau)$ denotes the derivative with respect to the second argument.

Similar results in the n centers case with milder requests of regularity for the $\alpha_i(t)$ and $y_i(t)$ can be proved for the corresponding parabolic problem. For details see [DFT1] and [DFT2].

We want finally consider a class of nonlinear evolution problems with point interaction concentrated in a single point (which for sake of simplicity will be taken to be the origin). The pointlike interaction models reproduce the basic features of the standard nonlinear Schrödinger equation in a simpler and more tractable context. Models of this kind, in space dimension one, appeared already in the physical literature ([J-LPS], [MA], [BKB]) to describe the resonant tunneling of electrons through a double barrier heterostructure.

Connections between standard nonlinear Schrödinger equation and point interaction Hamiltonians are more transparent in $d = 1$.

As an example, the one dimensional nonlinear Schrödinger equation with critical power nonlinearity is

$$i\frac{\partial \xi}{\partial t}(t,x) = -\frac{\partial^2 \xi}{\partial \xi^2}(t,x) - |\xi(t,x)|^4 \xi(t,x)$$

whereas the one dimensional Schrödinger equation with critical nonlinear delta interaction is given by

$$i\frac{\partial \eta}{\partial t}(t,x) = -\frac{\partial^2 \eta}{\partial \xi^2}(t,x) - |\eta(t,0)|^2 \delta_0 \eta(t,x) \tag{47}$$

Since for a class of blow-up solutions $\xi(t)$ of the first equation it has been shown that $|\xi(t)|^2 \to \delta_0$, it appears that, for this family of solutions, the second equation approximates the first one.

Unfortunately, this striking analogy does not hold in higher dimensions.

In the following we are going to consider the three dimensional case. For details on the one dimensional case see [AT1], [AT2].

Notice that even in the linear three dimensional case the solution of the evolution problem exhibits a singularity where the interaction is placed. This means that the nonlinear interaction cannot be introduced as in $d = 1$. The value of the function at the origin should be replaced in this case by the coefficient $q(t)$ of the singular part. This suggests the following formulation of the evolution problem (see (15))

$$\psi_t(x) = (U(t)\psi_0)(x) + i \int_0^t ds\, U(t-s, |x|) q(s)$$

$$q(t) + 4\sqrt{i\,\pi} \int_0^t ds \frac{\alpha(s)q(s)}{\sqrt{t-s}} = 4\sqrt{i\,\pi} \int_0^t ds \frac{(U(s)\psi_0)(0)}{\sqrt{t-s}}$$

$$\alpha(t) = \gamma |q(t)|^{2\sigma}, \quad \gamma \in \mathbb{R}, \ \sigma > 0 \tag{48}$$

The natural space where one looks for the solution of a standard nonlinear Schrödinger equation is H^1, i.e. the form domain of the generator of the linear dynamics obtained setting to zero the exponent of the nonlinear term. This suggests to look for the solution of (48) in the form domain $D(F_{\alpha,0})$ of $H_{\alpha,0}$. It is easy to check that the following characterization of $D(F_{\alpha,0})$ is equivalent to the one given in (17)

$$D(F_{\alpha,0}) = \{u \in L^2(\mathbb{R}^3) \mid u = \phi^\lambda + q G_\lambda, \ \phi^\lambda \in H^1(\mathbb{R}^3), \ q \in \mathbb{C}\}$$

One can prove the following result ([ADFT1])

Let $\psi_0 = \phi_0^\lambda + q G_\lambda \in D(F_{\alpha,0})$, with $\phi_0^\lambda \in H^2(\mathbb{R}^3)$ then there exists $\bar{t} > 0$ s.t. problem (48) has a unique solution $\psi_t \in D(F_{\alpha,0})$, $t \in [0, \bar{t})$. Moreover if either $\sigma < 1$ or $\gamma > 0$ the solution is global in time.

Global existence for repulsive interactions (i.e. $\gamma > 0$) is a direct consequences of the conservation of the L^2-norm and of the energy

$$E(\psi_t) = \|\nabla \phi_t\|^2 + \frac{\gamma}{\sigma + 1} |q(t)|^{2\sigma+2}$$

where $\psi_t = \phi_t + q(t)G_\lambda$.

One expects global existence also for weakly attractive interactions. The proof is not trivial because one is not working in H^1 and standard Sobolev inequalities are not available.

Consider finally the problem of the existence of blow-up solutions. A solution ψ_t of problem (48) will be said to be a blow-up solution if there exists a $t_0 < \infty$ such that

$$\lim_{t \to t_0} \|\nabla \phi_t\| = \infty$$

or equivalently

$$\lim_{t \to t_0} |q(t)| = \infty$$

With this definition of blow-up solutions, exploiting standard techniques in the theory of nonlinear (Schrödinger) equations it is possible to prove the following result ([ADFT2])

Let $\gamma < 0$, $\sigma \geq 1$ and $\psi_0 = \phi_0^\lambda + q_0 G_\lambda$, with $\phi_0^\lambda \in H^2(\mathbb{R}^3)$, $|x|\psi_0 \in L^2(\mathbb{R}^3)$, $E(\psi_0) < 0$.

Then the solution of problem (48) blows up in both directions of time.

References

[ABD]   Albeverio S., Brzeźniak Z., Dąbrowski L., *Fundamental Solution of the Heat and Schrödinger Equations with Point Interaction*, J. Func. Anal., **130**, 220–254, 1995.

[ADFT1]   Adami R., Dell'Antonio G., Figari R., Teta A., *The Cauchy Problem for the Schrödinger Equation in Dimension Three with Concentrated Nonlinearity*, Ann. Inst. H. Poincare' Non Lineaire, **20**, no. 3, 477-500, 2003.

[ADFT2]   Adami R., Dell'Antonio G., Figari R., Teta A., *Blow Up Solutions for the Schrödinger Equation with a Concentrated Nonlinearity in Dimension Three*, Ann. Inst. H. Poincare' Non Lineaire, **21**, no. 1, 121-137, 2004.

[AG]   Akhiezer N.I., Glazman I.M., *Theory of Linear Operators in Hilbert Space, Vol 2*, Pitman, Boston-London-Melbourne, 1981.

[AGH-KH]   Albeverio S., Gesztesy F., Högh-Krohn R., Holden H., *Solvable Models in Quantum Mechanics*, Springer-Verlag, New York, 1988.

[AT1]   Adami R., Teta A., *A Simple Model of Concentrated Nonlinearity*, in *Mathematical Results in Quantum Mechanics*, Dittrich J., Exner P.,Tater M. eds., Birkhäuser, 1999.

[AT2]   Adami R., Teta A., *A Class of Nonlinear Schrödinger Equation with Concentrated Nonlinearity*, J. Func. An., **180**,148-175, 2001.

[BFT]   Brasche J.F., Figari R., Teta A., *Singular Schrödinger Operator as Limits of Point Interaction Hamiltonians*, Potential Analysis, **8**, 163-178, 1998.

[BKB]   Bulashenko O.M., Kochelap V.A., Bonilla L.L., *Coherent Patterns and Self-Induced Diffraction of Electrons on a Thin Nonlinear Layer*, Phys. Rev. B, **54**, 3, 1996.

[DFT1]   Dell'Antonio G.F., Figari R., Teta A., *Diffusion of a Particle in Presence of N Moving Point Sources*, Ann. Inst. H. Poincare' A., **69**, 413-424, 1998.

[DFT2]   Dell'Antonio G.F., Figari R., Teta A., *Schrödinger Equation with Moving Point Interactions in Three Dimensions*, in *Stochastic Processes, Physics and Geometry: New Interplays, vol. I*, Gesztesy F., Holden H., Jost J., Paycha S., Röckner M., Scarlatti S. eds., A.M.S., Providence, 2000.

[DG]   Dąbrowski L., Grosse H., *On nonlocal point interactions in one, two, and three dimensions*, J. Math. Phys., **26**, 2777-2780, 1985.

[F]   Fermi E., *Sul moto dei neutroni nelle sostanze idrogenate*, Ricerca Scientifica **7**, 13-52, 1936 (Engl. Trans. in *E. Fermi Collected Papers Vol.1*, University of Chicago, Chicago-London, 1962).

[FHT]   Figari R., Holden H., Teta A., *A law of large numbers and a central limit theorem for the Schrödinger operator with zero range potentials*, J. Stat. Phys., **51**, 205-214, 1988.

[FT]   Figari R., Teta A., *A boundary value problem of mixed type on perforated domains* Asymptotic Analysis, **6**, 271-284, 1993.

[J-LPS]   Jona-Lasinio G., Presilla C., Sjöstrand J., *On Schrödinger Equations with Concentrated Nonlinearities*, Ann. Phys., **240**, 1-21, 1995.

[KP]   Kronig R. de L., Penney W. G., *Quantum Mechanics of Electrons in Crystal Lattices*, Proc. Roy. Soc. (London), **130A**, 499-513, 1931.

[MA]   Malomed B., Azbel M., *Modulational Instability of a Wave Scattered by a Nonlinear Center*, Phys. Rev. B, **47**, 16, 1993.

[S]   Schulman L.S., *Application of the propagator for the delta function potential*, in *Path integrals from mev to Mev*, Gutzwiller M.C., Iuomata A., Klauder J.K., Streit L. eds., World Scientific, Singapore, 1986.

[ST]   Scarlatti S., Teta A., *Derivation of the Time-dependent Propagator for the Three Dimensional Schrödinger Equation with One Point Interaction*, J. Phys. A, **23**, 1033-1035, 1990.

[Si]   Simon B., *Quantum Mechanics for Hamiltonians Defined as Quadratic Forms*, Princeton University Press, 1971.

[SY]   Sayapova M.R., Yafaev D.R., *The evolution operator for time-dependent potentials of zero radius*, Proc. Stek. Inst. Math., **2**, 173-180, 1984.

[T]   Teta A., *Quadratic Forms for Singular Perturbations of the Laplacian*, Publications of the R.I.M.S., Kyoto University, **26**, 803-817, 1990.

List of Participants

1. Antonelli Laura
   University of Napoli, Italy
   `laura.antonelli@dma.unina.it`
2. Arridge Simon R.
   University College London
   `S.Arridge@cs.ucl.ac.uk`
3. Bandiera Francesco
   University of Lecce, Italy
   `francesco.bandiera@unile.it`
4. Berowski Przemyslaw
   Warsaw University, Polonia
   `berowski@iel.waw.pl`
5. Bonilla Luis
   University of Carlos III de Madrid,
   Spain
   (**editor**)
6. Canfora Michela
   University of Firenze, Italy
   `canfora@ge.infm.it`
7. Carpio Anna
   University of Complutense de
   Madrid, Spain
   `ana_carpio@mat.ucm.es`
8. Cebrian Elena
   University of Burgos, Spain
   `elenac@ubu.es`
9. Chukalina Marina
   Institute of Microelectronics,
   Moscow, Russia
   `marina@ipmt-hpm.ac.ru`
10. Cuoghi Paola
    University of Modena, Italy
    `pcuoghi@unimo.it`
11. Dell'Antonio Gianfausto
    University of Roma1, Italy
    `dellantonio@mat.uniroma1.it`
12. Di Cristo Michele
    University of Milano, Italy
    `dicristo@mat.unimi.it`
13. Dorn Oliver
    University of Carlos III de Madrid,
    Spain
    `dorn@cs.ubc.ca`
    (**lecturer**)
14. Dubovskii Pavel
    Russian Academy of Sciences,
    Russia
    `dubovski@inm.ras.ru`
15. Eelbode David
    University of Ghent, Belgium
    `deef@cage.rug.ac.be`
16. Galletti Ardelio
    University of Napoli, Italy
    `ardelio.galletti@email.it`
17. Gonzalez Pedro
    University of Carlos III de Madrid,
    Spain
    `pgonzale@ing.uc3m.es`

18. Gosse Laurent
    IAMI-CNR, Bari, Italy
    `l.gosse@area.ba.cnr.it`
19. Hristova Veronica
    University of Sofia, Bulgaria
    `veronica_hristova@hotmail.com`
20. Landi Germana
    University of Bologna, Italy
    `landig@dm.unibo.it`
21. Leitao Antonio
    University of Santa Catarina,
    Brasil
    `aleitao@mtm.ufsc.br`
22. Luquin Brigitte
    University Pierre et M. Curie,
    France
    `lucquin@ann.jussieu.fr`
23. Makarenkov Oleg
    University of Voronezh, Russia
    `omakarenkov@mail.ru`
24. Makrakis George
    University of Creta, Greece
    `makrakg@iacm.forth.gr`
25. Mallaina Eduardo Fed
    University of Buenos Aires,
    Argentina
    `efmalla@fi.uba.ar`
26. Mariani Francesca
    University of Firenze, Italy
    `mariani@math.unifi.it`
27. Marino Zelda
    University of Napoli, Italy
    `zelda_marino@dma.unina.it`
28. Massone Anna Maria
    INFM Genova, Italy
    `massone@ge.infm.it`
29. Mastronardi Nicola
    CNR, Bari, Italy
    `n.mastronardi@area.ba.cnr.it`
30. Moscoso Miguel
    University Carlos III de Madrid,
    Spain
    `moscoso@math.uc3m.es`
    (**lecturer**)

31. Natterer Frank
    Westfälische Wilhelms-University
    Münster, Germany
    `nattere@math.uni-muenster.de`
    (**lecturer**)
32. Orlandi Enza
    University of Roma3, Italy
    `orlandi@matrm3.mat.uniroma3.it`
33. Papanicolaou George
    Stanford University, USA
    `papanico@math.stanford.edu`
34. Rebolledo Aldo Franco
    University of del Valle, Colombia
    `aldoptica@hotmail.com`
35. Rey Fernando Gonzalo
    University of Buenos Aires,
    Argentina
    `hrey@fi.uba.ar`
36. Salani Claudia
    University of Milano, Italy
    `salani@mat.unimi.it`
37. Sanchez Oscar
    University of Granada, Spain
    `ossanche@ugr.es`
38. Skaug Christian
    CNR, Bari, Italy
    `c.skaug@area.ba.cnr.it`
39. Soleimani Manuchehr
    University of Manhester, UK
    `m.soleimani@student.umist.ac.uk`
40. Stasiak Magdalena
    University of Lodz, Polonia
    `stasiak@ck-sg.p.lodz.pl`
41. Symes William S.
    Rice University, USA
    `symes@caam.rice.edu`
42. Teta Alessandro
    University of L' Aquila, Italy
    `teta@univaq.it`
    (**lecturer**)
43. Zama Fabiana
    University of Bologna, Italy
    `zama@dm.unibo.it`

LIST OF C.I.M.E. SEMINARS

Published by C.I.M.E

1954	1. Analisi funzionale
	2. Quadratura delle superficie e questioni connesse
	3. Equazioni differenziali non lineari
1955	4. Teorema di Riemann-Roch e questioni connesse
	5. Teoria dei numeri
	6. Topologia
	7. Teorie non linearizzate in elasticità, idrodinamica, aerodinamic
	8. Geometria proiettivo-differenziale
1956	9. Equazioni alle derivate parziali a caratteristiche reali
	10. Propagazione delle onde elettromagnetiche automorfe
	11. Teoria della funzioni di più variabili complesse e delle funzioni
1957	12. Geometria aritmetica e algebrica (2 vol.)
	13. Integrali singolari e questioni connesse
	14. Teoria della turbolenza (2 vol.)
1958	15. Vedute e problemi attuali in relatività generale
	16. Problemi di geometria differenziale in grande
	17. Il principio di minimo e le sue applicazioni alle equazioni funzionali
1959	18. Induzione e statistica
	19. Teoria algebrica dei meccanismi automatici (2 vol.)
	20. Gruppi, anelli di Lie e teoria della coomologia
1960	21. Sistemi dinamici e teoremi ergodici
	22. Forme differenziali e loro integrali
1961	23. Geometria del calcolo delle variazioni (2 vol.)
	24. Teoria delle distribuzioni
	25. Onde superficiali
1962	26. Topologia differenziale
	27. Autovalori e autosoluzioni
	28. Magnetofluidodinamica
1963	29. Equazioni differenziali astratte
	30. Funzioni e varietà complesse
	31. Proprietà di media e teoremi di confronto in Fisica Matematica
1964	32. Relatività generale
	33. Dinamica dei gas rarefatti
	34. Alcune questioni di analisi numerica
	35. Equazioni differenziali non lineari
1965	36. Non-linear continuum theories
	37. Some aspects of ring theory
	38. Mathematical optimization in economics

Published by Ed. Cremonese, Firenze

1966   39. Calculus of variations
             40. Economia matematica
             41. Classi caratteristiche e questioni connesse
             42. Some aspects of diffusion theory

1967   43. Modern questions of celestial mechanics
             44. Numerical analysis of partial differential equations
             45. Geometry of homogeneous bounded domains

1968   46. Controllability and observability
             47. Pseudo-differential operators
             48. Aspects of mathematical logic

1969   49. Potential theory
             50. Non-linear continuum theories in mechanics and physics and their applications
             51. Questions of algebraic varieties

1970   52. Relativistic fluid dynamics
             53. Theory of group representations and Fourier analysis
             54. Functional equations and inequalities
             55. Problems in non-linear analysis

1971   56. Stereodynamics
             57. Constructive aspects of functional analysis (2 vol.)
             58. Categories and commutative algebra

1972   59. Non-linear mechanics
             60. Finite geometric structures and their applications
             61. Geometric measure theory and minimal surfaces

1973   62. Complex analysis
             63. New variational techniques in mathematical physics
             64. Spectral analysis

1974   65. Stability problems
             66. Singularities of analytic spaces
             67. Eigenvalues of non linear problems

1975   68. Theoretical computer sciences
             69. Model theory and applications
             70. Differential operators and manifolds

Published by Ed. Liguori, Napoli

1976   71. Statistical Mechanics
             72. Hyperbolicity
             73. Differential topology

1977   74. Materials with memory
             75. Pseudodifferential operators with applications
             76. Algebraic surfaces

Published by Ed. Liguori, Napoli & Birkhäuser

1978   77. Stochastic differential equations
             78. Dynamical systems

1979   79. Recursion theory and computational complexity
             80. Mathematics of biology

1980   81. Wave propagation
             82. Harmonic analysis and group representations
             83. Matroid theory and its applications

Published by Springer-Verlag

1981   84. Kinetic Theories and the Boltzmann Equation   (LNM 1048)
      85. Algebraic Threefolds   (LNM 947)
      86. Nonlinear Filtering and Stochastic Control   (LNM 972)

1982   87. Invariant Theory   (LNM 996)
      88. Thermodynamics and Constitutive Equations   (LNP 228)
      89. Fluid Dynamics   (LNM 1047)

1983   90. Complete Intersections   (LNM 1092)
      91. Bifurcation Theory and Applications   (LNM 1057)
      92. Numerical Methods in Fluid Dynamics   (LNM 1127)

1984   93. Harmonic Mappings and Minimal Immersions   (LNM 1161)
      94. Schrödinger Operators   (LNM 1159)
      95. Buildings and the Geometry of Diagrams   (LNM 1181)

1985   96. Probability and Analysis   (LNM 1206)
      97. Some Problems in Nonlinear Diffusion   (LNM 1224)
      98. Theory of Moduli   (LNM 1337)

1986   99. Inverse Problems   (LNM 1225)
    100. Mathematical Economics   (LNM 1330)
    101. Combinatorial Optimization   (LNM 1403)

1987   102. Relativistic Fluid Dynamics   (LNM 1385)
    103. Topics in Calculus of Variations   (LNM 1365)

1988   104. Logic and Computer Science   (LNM 1429)
    105. Global Geometry and Mathematical Physics   (LNM 1451)

1989   106. Methods of nonconvex analysis   (LNM 1446)
    107. Microlocal Analysis and Applications   (LNM 1495)

1990   108. Geometric Topology: Recent Developments   (LNM 1504)
    109. H_∞ Control Theory   (LNM 1496)
    110. Mathematical Modelling of Industrial Processes   (LNM 1521)

1991   111. Topological Methods for Ordinary Differential Equations   (LNM 1537)
    112. Arithmetic Algebraic Geometry   (LNM 1553)
    113. Transition to Chaos in Classical and Quantum Mechanics   (LNM 1589)

1992   114. Dirichlet Forms   (LNM 1563)
    115. D-Modules, Representation Theory, and Quantum Groups   (LNM 1565)
    116. Nonequilibrium Problems in Many-Particle Systems   (LNM 1551)

1993   117. Integrable Systems and Quantum Groups   (LNM 1620)
    118. Algebraic Cycles and Hodge Theory   (LNM 1594)
    119. Phase Transitions and Hysteresis   (LNM 1584)

1994   120. Recent Mathematical Methods in Nonlinear Wave Propagation   (LNM 1640)
    121. Dynamical Systems   (LNM 1609)
    122. Transcendental Methods in Algebraic Geometry   (LNM 1646)

1995   123. Probabilistic Models for Nonlinear PDE's   (LNM 1627)
    124. Viscosity Solutions and Applications   (LNM 1660)
    125. Vector Bundles on Curves. New Directions   (LNM 1649)

1996   126. Integral Geometry, Radon Transforms and Complex Analysis   (LNM 1684)
    127. Calculus of Variations and Geometric Evolution Problems   (LNM 1713)
    128. Financial Mathematics   (LNM 1656)

1997	129. Mathematics Inspired by Biology	(LNM 1714)
	130. Advanced Numerical Approximation of Nonlinear Hyperbolic Equations	(LNM 1697)
	131. Arithmetic Theory of Elliptic Curves	(LNM 1716)
	132. Quantum Cohomology	(LNM 1776)
1998	133. Optimal Shape Design	(LNM 1740)
	134. Dynamical Systems and Small Divisors	(LNM 1784)
	135. Mathematical Problems in Semiconductor Physics	(LNM 1823)
	136. Stochastic PDE's and Kolmogorov Equations in Infinite Dimension	(LNM 1715)
	137. Filtration in Porous Media and Industrial Applications	(LNM 1734)
1999	138. Computational Mathematics driven by Industrial Applications	(LNM 1739)
	139. Iwahori-Hecke Algebras and Representation Theory	(LNM 1804)
	140. Hamiltonian Dynamics - Theory and Applications	(LNM 1861)
	141. Global Theory of Minimal Surfaces in Flat Spaces	(LNM 1775)
	142. Direct and Inverse Methods in Solving Nonlinear Evolution Equations	(LNP 632)
2000	143. Dynamical Systems	(LNM 1822)
	144. Diophantine Approximation	(LNM 1819)
	145. Mathematical Aspects of Evolving Interfaces	(LNM 1812)
	146. Mathematical Methods for Protein Structure	(LNCS 2666)
	147. Noncommutative Geometry	(LNM 1831)
2001	148. Topological Fluid Mechanics	to appear
	149. Spatial Stochastic Processes	(LNM 1802)
	150. Optimal Transportation and Applications	(LNM 1813)
	151. Multiscale Problems and Methods in Numerical Simulations	(LNM 1825)
2002	152. Real Methods in Complex and CR Geometry	(LNM 1848)
	153. Analytic Number Theory	(LNM 1891)
	154. Inverse Problems and Imaging	(LNM 1943)
2003	155. Stochastic Methods in Finance	(LNM 1856)
	156. Hyperbolic Systems of Balance Laws	(LNM 1911)
	157. Symplectic 4-Manifolds and Algebraic Surfaces	(LNM 1938)
	158. Mathematical Foundation of Turbulent Viscous Flows	(LNM 1871)
2004	159. Representation Theory and Complex Analysis	(LNM 1931)
	160. Nonlinear and Optimal Control Theory	(LNM 1932)
	161. Stochastic Geometry	(LNM 1892)
2005	162. Enumerative Invariants in Algebraic Geometry and String Theory	to appear
	163. Calculus of Variations and Non-linear Partial Differential Equations	(LNM 1927)
	164. SPDE in Hydrodynamic. Recent Progress and Prospects	(LNM 1942)
2006	165. Pseudo-Differential Operators, Quantization and Signals	to appear
	166. Mixed Finite Elements, Compatibility Conditions, and Applications	(LNM 1939)
	167. Multiscale Problems in the Life Sciences. From Microscopic to Macroscopic	(LNM 1940)
	168. Quantum Transport: Modelling, Analysis and Asymptotics	to appear
2007	169. Geometric Analysis and Partial Differential Equations	to appear
	170. Nonlinear Optimization	to appear
	171. Arithmetic Geometry	to appear
2008	172. Nonlinear Partial Differential Equations and Applications	announced
	173. Holomorphic Dynamical Systems	announced
	174. Level Set and PDE based Reconstruction Methods: Applications to Inverse Problems and Image Processing	announced
	175. Mathematical models in the manufacturing of glass, polymers and textiles	announced

Lecture Notes in Mathematics

For information about earlier volumes
please contact your bookseller or Springer
LNM Online archive: springerlink.com

Vol. 1755: J. Azéma, M. Émery, M. Ledoux, M. Yor (Eds.), Séminaire de Probabilités XXXV (2001)

Vol. 1756: P. E. Zhidkov, Korteweg de Vries and Nonlinear Schrödinger Equations: Qualitative Theory (2001)

Vol. 1757: R. R. Phelps, Lectures on Choquet's Theorem (2001)

Vol. 1758: N. Monod, Continuous Bounded Cohomology of Locally Compact Groups (2001)

Vol. 1759: Y. Abe, K. Kopfermann, Toroidal Groups (2001)

Vol. 1760: D. Filipović, Consistency Problems for Heath-Jarrow-Morton Interest Rate Models (2001)

Vol. 1761: C. Adelmann, The Decomposition of Primes in Torsion Point Fields (2001)

Vol. 1762: S. Cerrai, Second Order PDE's in Finite and Infinite Dimension (2001)

Vol. 1763: J.-L. Loday, A. Frabetti, F. Chapoton, F. Goichot, Dialgebras and Related Operads (2001)

Vol. 1764: A. Cannas da Silva, Lectures on Symplectic Geometry (2001)

Vol. 1765: T. Kerler, V. V. Lyubashenko, Non-Semisimple Topological Quantum Field Theories for 3-Manifolds with Corners (2001)

Vol. 1766: H. Hennion, L. Hervé, Limit Theorems for Markov Chains and Stochastic Properties of Dynamical Systems by Quasi-Compactness (2001)

Vol. 1767: J. Xiao, Holomorphic Q Classes (2001)

Vol. 1768: M. J. Pflaum, Analytic and Geometric Study of Stratified Spaces (2001)

Vol. 1769: M. Alberich-Carramiñana, Geometry of the Plane Cremona Maps (2002)

Vol. 1770: H. Gluesing-Luerssen, Linear Delay-Differential Systems with Commensurate Delays: An Algebraic Approach (2002)

Vol. 1771: M. Émery, M. Yor (Eds.), Séminaire de Probabilités 1967-1980. A Selection in Martingale Theory (2002)

Vol. 1772: F. Burstall, D. Ferus, K. Leschke, F. Pedit, U. Pinkall, Conformal Geometry of Surfaces in S^4 (2002)

Vol. 1773: Z. Arad, M. Muzychuk, Standard Integral Table Algebras Generated by a Non-real Element of Small Degree (2002)

Vol. 1774: V. Runde, Lectures on Amenability (2002)

Vol. 1775: W. H. Meeks, A. Ros, H. Rosenberg, The Global Theory of Minimal Surfaces in Flat Spaces. Martina Franca 1999. Editor: G. P. Pirola (2002)

Vol. 1776: K. Behrend, C. Gomez, V. Tarasov, G. Tian, Quantum Comohology. Cetraro 1997. Editors: P. de Bartolomeis, B. Dubrovin, C. Reina (2002)

Vol. 1777: E. García-Río, D. N. Kupeli, R. Vázquez-Lorenzo, Osserman Manifolds in Semi-Riemannian Geometry (2002)

Vol. 1778: H. Kiechle, Theory of K-Loops (2002)

Vol. 1779: I. Chueshov, Monotone Random Systems (2002)

Vol. 1780: J. H. Bruinier, Borcherds Products on O(2,1) and Chern Classes of Heegner Divisors (2002)

Vol. 1781: E. Bolthausen, E. Perkins, A. van der Vaart, Lectures on Probability Theory and Statistics. Ecole d' Eté de Probabilités de Saint-Flour XXIX-1999. Editor: P. Bernard (2002)

Vol. 1782: C.-H. Chu, A. T.-M. Lau, Harmonic Functions on Groups and Fourier Algebras (2002)

Vol. 1783: L. Grüne, Asymptotic Behavior of Dynamical and Control Systems under Perturbation and Discretization (2002)

Vol. 1784: L. H. Eliasson, S. B. Kuksin, S. Marmi, J.-C. Yoccoz, Dynamical Systems and Small Divisors. Cetraro, Italy 1998. Editors: S. Marmi, J.-C. Yoccoz (2002)

Vol. 1785: J. Arias de Reyna, Pointwise Convergence of Fourier Series (2002)

Vol. 1786: S. D. Cutkosky, Monomialization of Morphisms from 3-Folds to Surfaces (2002)

Vol. 1787: S. Caenepeel, G. Militaru, S. Zhu, Frobenius and Separable Functors for Generalized Module Categories and Nonlinear Equations (2002)

Vol. 1788: A. Vasil'ev, Moduli of Families of Curves for Conformal and Quasiconformal Mappings (2002)

Vol. 1789: Y. Sommerhäuser, Yetter-Drinfel'd Hopf algebras over groups of prime order (2002)

Vol. 1790: X. Zhan, Matrix Inequalities (2002)

Vol. 1791: M. Knebusch, D. Zhang, Manis Valuations and Prüfer Extensions I: A new Chapter in Commutative Algebra (2002)

Vol. 1792: D. D. Ang, R. Gorenflo, V. K. Le, D. D. Trong, Moment Theory and Some Inverse Problems in Potential Theory and Heat Conduction (2002)

Vol. 1793: J. Cortés Monforte, Geometric, Control and Numerical Aspects of Nonholonomic Systems (2002)

Vol. 1794: N. Pytheas Fogg, Substitution in Dynamics, Arithmetics and Combinatorics. Editors: V. Berthé, S. Ferenczi, C. Mauduit, A. Siegel (2002)

Vol. 1795: H. Li, Filtered-Graded Transfer in Using Noncommutative Gröbner Bases (2002)

Vol. 1796: J.M. Melenk, hp-Finite Element Methods for Singular Perturbations (2002)

Vol. 1797: B. Schmidt, Characters and Cyclotomic Fields in Finite Geometry (2002)

Vol. 1798: W.M. Oliva, Geometric Mechanics (2002)

Vol. 1799: H. Pajot, Analytic Capacity, Rectifiability, Menger Curvature and the Cauchy Integral (2002)

Vol. 1800: O. Gabber, L. Ramero, Almost Ring Theory (2003)

Vol. 1801: J. Azéma, M. Émery, M. Ledoux, M. Yor (Eds.), Séminaire de Probabilités XXXVI (2003)

Vol. 1802: V. Capasso, E. Merzbach, B. G. Ivanoff, M. Dozzi, R. Dalang, T. Mountford, Topics in Spatial Stochastic Processes. Martina Franca, Italy 2001. Editor: E. Merzbach (2003)

Vol. 1803: G. Dolzmann, Variational Methods for Crystalline Microstructure – Analysis and Computation (2003)

Vol. 1804: I. Cherednik, Ya. Markov, R. Howe, G. Lusztig, Iwahori-Hecke Algebras and their Representation Theory. Martina Franca, Italy 1999. Editors: V. Baldoni, D. Barbasch (2003)

Vol. 1805: F. Cao, Geometric Curve Evolution and Image Processing (2003)

Vol. 1806: H. Broer, I. Hoveijn. G. Lunther, G. Vegter, Bifurcations in Hamiltonian Systems. Computing Singularities by Gröbner Bases (2003)

Vol. 1807: V. D. Milman, G. Schechtman (Eds.), Geometric Aspects of Functional Analysis. Israel Seminar 2000-2002 (2003)

Vol. 1808: W. Schindler, Measures with Symmetry Properties (2003)

Vol. 1809: O. Steinbach, Stability Estimates for Hybrid Coupled Domain Decomposition Methods (2003)

Vol. 1810: J. Wengenroth, Derived Functors in Functional Analysis (2003)

Vol. 1811: J. Stevens, Deformations of Singularities (2003)

Vol. 1812: L. Ambrosio, K. Deckelnick, G. Dziuk, M. Mimura, V. A. Solonnikov, H. M. Soner, Mathematical Aspects of Evolving Interfaces. Madeira, Funchal, Portugal 2000. Editors: P. Colli, J. F. Rodrigues (2003)

Vol. 1813: L. Ambrosio, L. A. Caffarelli, Y. Brenier, G. Buttazzo, C. Villani, Optimal Transportation and its Applications. Martina Franca, Italy 2001. Editors: L. A. Caffarelli, S. Salsa (2003)

Vol. 1814: P. Bank, F. Baudoin, H. Föllmer, L.C.G. Rogers, M. Soner, N. Touzi, Paris-Princeton Lectures on Mathematical Finance 2002 (2003)

Vol. 1815: A. M. Vershik (Ed.), Asymptotic Combinatorics with Applications to Mathematical Physics. St. Petersburg, Russia 2001 (2003)

Vol. 1816: S. Albeverio, W. Schachermayer, M. Talagrand, Lectures on Probability Theory and Statistics. Ecole d'Eté de Probabilités de Saint-Flour XXX-2000. Editor: P. Bernard (2003)

Vol. 1817: E. Koelink, W. Van Assche (Eds.), Orthogonal Polynomials and Special Functions. Leuven 2002 (2003)

Vol. 1818: M. Bildhauer, Convex Variational Problems with Linear, nearly Linear and/or Anisotropic Growth Conditions (2003)

Vol. 1819: D. Masser, Yu. V. Nesterenko, H. P. Schlickewei, W. M. Schmidt, M. Waldschmidt, Diophantine Approximation. Cetraro, Italy 2000. Editors: F. Amoroso, U. Zannier (2003)

Vol. 1820: F. Hiai, H. Kosaki, Means of Hilbert Space Operators (2003)

Vol. 1821: S. Teufel, Adiabatic Perturbation Theory in Quantum Dynamics (2003)

Vol. 1822: S.-N. Chow, R. Conti, R. Johnson, J. Mallet-Paret, R. Nussbaum, Dynamical Systems. Cetraro, Italy 2000. Editors: J. W. Macki, P. Zecca (2003)

Vol. 1823: A. M. Anile, W. Allegretto, C. Ringhofer, Mathematical Problems in Semiconductor Physics. Cetraro, Italy 1998. Editor: A. M. Anile (2003)

Vol. 1824: J. A. Navarro González, J. B. Sancho de Salas, $\mathscr{C}^\infty$ – Differentiable Spaces (2003)

Vol. 1825: J. H. Bramble, A. Cohen, W. Dahmen, Multiscale Problems and Methods in Numerical Simulations, Martina Franca, Italy 2001. Editor: C. Canuto (2003)

Vol. 1826: K. Dohmen, Improved Bonferroni Inequalities via Abstract Tubes. Inequalities and Identities of Inclusion-Exclusion Type. VIII, 113 p, 2003.

Vol. 1827: K. M. Pilgrim, Combinations of Complex Dynamical Systems. IX, 118 p, 2003.

Vol. 1828: D. J. Green, Gröbner Bases and the Computation of Group Cohomology. XII, 138 p, 2003.

Vol. 1829: E. Altman, B. Gaujal, A. Hordijk, Discrete-Event Control of Stochastic Networks: Multimodularity and Regularity. XIV, 313 p, 2003.

Vol. 1830: M. I. Gil', Operator Functions and Localization of Spectra. XIV, 256 p, 2003.

Vol. 1831: A. Connes, J. Cuntz, E. Guentner, N. Higson, J. E. Kaminker, Noncommutative Geometry, Martina Franca, Italy 2002. Editors: S. Doplicher, L. Longo (2004)

Vol. 1832: J. Azéma, M. Émery, M. Ledoux, M. Yor (Eds.), Séminaire de Probabilités XXXVII (2003)

Vol. 1833: D.-Q. Jiang, M. Qian, M.-P. Qian, Mathematical Theory of Nonequilibrium Steady States. On the Frontier of Probability and Dynamical Systems. IX, 280 p, 2004.

Vol. 1834: Yo. Yomdin, G. Comte, Tame Geometry with Application in Smooth Analysis. VIII, 186 p, 2004.

Vol. 1835: O.T. Izhboldin, B. Kahn, N.A. Karpenko, A. Vishik, Geometric Methods in the Algebraic Theory of Quadratic Forms. Summer School, Lens, 2000. Editor: J.-P. Tignol (2004)

Vol. 1836: C. Năstăsescu, F. Van Oystaeyen, Methods of Graded Rings. XIII, 304 p, 2004.

Vol. 1837: S. Tavaré, O. Zeitouni, Lectures on Probability Theory and Statistics. Ecole d'Eté de Probabilités de Saint-Flour XXXI-2001. Editor: J. Picard (2004)

Vol. 1838: A.J. Ganesh, N.W. O'Connell, D.J. Wischik, Big Queues. XII, 254 p, 2004.

Vol. 1839: R. Gohm, Noncommutative Stationary Processes. VIII, 170 p, 2004.

Vol. 1840: B. Tsirelson, W. Werner, Lectures on Probability Theory and Statistics. Ecole d'Eté de Probabilités de Saint-Flour XXXII-2002. Editor: J. Picard (2004)

Vol. 1841: W. Reichel, Uniqueness Theorems for Variational Problems by the Method of Transformation Groups (2004)

Vol. 1842: T. Johnsen, A. L. Knutsen, K_3 Projective Models in Scrolls (2004)

Vol. 1843: B. Jefferies, Spectral Properties of Noncommuting Operators (2004)

Vol. 1844: K.F. Siburg, The Principle of Least Action in Geometry and Dynamics (2004)

Vol. 1845: Min Ho Lee, Mixed Automorphic Forms, Torus Bundles, and Jacobi Forms (2004)

Vol. 1846: H. Ammari, H. Kang, Reconstruction of Small Inhomogeneities from Boundary Measurements (2004)

Vol. 1847: T.R. Bielecki, T. Björk, M. Jeanblanc, M. Rutkowski, J.A. Scheinkman, W. Xiong, Paris-Princeton Lectures on Mathematical Finance 2003 (2004)

Vol. 1848: M. Abate, J. E. Fornaess, X. Huang, J. P. Rosay, A. Tumanov, Real Methods in Complex and CR Geometry, Martina Franca, Italy 2002. Editors: D. Zaitsev, G. Zampieri (2004)

Vol. 1849: Martin L. Brown, Heegner Modules and Elliptic Curves (2004)

Vol. 1850: V. D. Milman, G. Schechtman (Eds.), Geometric Aspects of Functional Analysis. Israel Seminar 2002-2003 (2004)

Vol. 1851: O. Catoni, Statistical Learning Theory and Stochastic Optimization (2004)

Vol. 1852: A.S. Kechris, B.D. Miller, Topics in Orbit Equivalence (2004)

Vol. 1853: Ch. Favre, M. Jonsson, The Valuative Tree (2004)

Vol. 1854: O. Saeki, Topology of Singular Fibers of Differential Maps (2004)

Vol. 1855: G. Da Prato, P.C. Kunstmann, I. Lasiecka, A. Lunardi, R. Schnaubelt, L. Weis, Functional Analytic Methods for Evolution Equations. Editors: M. Iannelli, R. Nagel, S. Piazzera (2004)

Vol. 1856: K. Back, T.R. Bielecki, C. Hipp, S. Peng, W. Schachermayer, Stochastic Methods in Finance, Bressanone/Brixen, Italy, 2003. Editors: M. Fritelli, W. Runggaldier (2004)

Vol. 1857: M. Émery, M. Ledoux, M. Yor (Eds.), Séminaire de Probabilités XXXVIII (2005)

Vol. 1858: A.S. Cherny, H.-J. Engelbert, Singular Stochastic Differential Equations (2005)

Vol. 1859: E. Letellier, Fourier Transforms of Invariant Functions on Finite Reductive Lie Algebras (2005)

Vol. 1860: A. Borisyuk, G.B. Ermentrout, A. Friedman, D. Terman, Tutorials in Mathematical Biosciences I. Mathematical Neurosciences (2005)

Vol. 1861: G. Benettin, J. Henrard, S. Kuksin, Hamiltonian Dynamics – Theory and Applications, Cetraro, Italy, 1999. Editor: A. Giorgilli (2005)

Vol. 1862: B. Helffer, F. Nier, Hypoelliptic Estimates and Spectral Theory for Fokker-Planck Operators and Witten Laplacians (2005)

Vol. 1863: H. Führ, Abstract Harmonic Analysis of Continuous Wavelet Transforms (2005)

Vol. 1864: K. Efstathiou, Metamorphoses of Hamiltonian Systems with Symmetries (2005)

Vol. 1865: D. Applebaum, B.V. R. Bhat, J. Kustermans, J. M. Lindsay, Quantum Independent Increment Processes I. From Classical Probability to Quantum Stochastic Calculus. Editors: M. Schürmann, U. Franz (2005)

Vol. 1866: O.E. Barndorff-Nielsen, U. Franz, R. Gohm, B. Kümmerer, S. Thorbjønsen, Quantum Independent Increment Processes II. Structure of Quantum Lévy Processes, Classical Probability, and Physics. Editors: M. Schürmann, U. Franz, (2005)

Vol. 1867: J. Sneyd (Ed.), Tutorials in Mathematical Biosciences II. Mathematical Modeling of Calcium Dynamics and Signal Transduction. (2005)

Vol. 1868: J. Jorgenson, S. Lang, $Pos_n(R)$ and Eisenstein Series. (2005)

Vol. 1869: A. Dembo, T. Funaki, Lectures on Probability Theory and Statistics. Ecole d'Eté de Probabilités de Saint-Flour XXXIII-2003. Editor: J. Picard (2005)

Vol. 1870: V.I. Gurariy, W. Lusky, Geometry of Müntz Spaces and Related Questions. (2005)

Vol. 1871: P. Constantin, G. Gallavotti, A.V. Kazhikhov, Y. Meyer, S. Ukai, Mathematical Foundation of Turbulent Viscous Flows, Martina Franca, Italy, 2003. Editors: M. Cannone, T. Miyakawa (2006)

Vol. 1872: A. Friedman (Ed.), Tutorials in Mathematical Biosciences III. Cell Cycle, Proliferation, and Cancer (2006)

Vol. 1873: R. Mansuy, M. Yor, Random Times and Enlargements of Filtrations in a Brownian Setting (2006)

Vol. 1874: M. Yor, M. Émery (Eds.), In Memoriam Paul-André Meyer - Séminaire de Probabilités XXXIX (2006)

Vol. 1875: J. Pitman, Combinatorial Stochastic Processes. Ecole d'Eté de Probabilités de Saint-Flour XXXII-2002. Editor: J. Picard (2006)

Vol. 1876: H. Herrlich, Axiom of Choice (2006)

Vol. 1877: J. Steuding, Value Distributions of L-Functions (2007)

Vol. 1878: R. Cerf, The Wulff Crystal in Ising and Percolation Models, Ecole d'Eté de Probabilités de Saint-Flour XXXIV-2004. Editor: Jean Picard (2006)

Vol. 1879: G. Slade, The Lace Expansion and its Applications, Ecole d'Eté de Probabilités de Saint-Flour XXXIV-2004. Editor: Jean Picard (2006)

Vol. 1880: S. Attal, A. Joye, C.-A. Pillet, Open Quantum Systems I, The Hamiltonian Approach (2006)

Vol. 1881: S. Attal, A. Joye, C.-A. Pillet, Open Quantum Systems II, The Markovian Approach (2006)

Vol. 1882: S. Attal, A. Joye, C.-A. Pillet, Open Quantum Systems III, Recent Developments (2006)

Vol. 1883: W. Van Assche, F. Marcellàn (Eds.), Orthogonal Polynomials and Special Functions, Computation and Application (2006)

Vol. 1884: N. Hayashi, E.I. Kaikina, P.I. Naumkin, I.A. Shishmarev, Asymptotics for Dissipative Nonlinear Equations (2006)

Vol. 1885: A. Telcs, The Art of Random Walks (2006)

Vol. 1886: S. Takamura, Splitting Deformations of Degenerations of Complex Curves (2006)

Vol. 1887: K. Habermann, L. Habermann, Introduction to Symplectic Dirac Operators (2006)

Vol. 1888: J. van der Hoeven, Transseries and Real Differential Algebra (2006)

Vol. 1889: G. Osipenko, Dynamical Systems, Graphs, and Algorithms (2006)

Vol. 1890: M. Bunge, J. Funk, Singular Coverings of Toposes (2006)

Vol. 1891: J.B. Friedlander, D.R. Heath-Brown, H. Iwaniec, J. Kaczorowski, Analytic Number Theory, Cetraro, Italy, 2002. Editors: A. Perelli, C. Viola (2006)

Vol. 1892: A. Baddeley, I. Bárány, R. Schneider, W. Weil, Stochastic Geometry, Martina Franca, Italy, 2004. Editor: W. Weil (2007)

Vol. 1893: H. Hanßmann, Local and Semi-Local Bifurcations in Hamiltonian Dynamical Systems, Results and Examples (2007)

Vol. 1894: C.W. Groetsch, Stable Approximate Evaluation of Unbounded Operators (2007)

Vol. 1895: L. Molnár, Selected Preserver Problems on Algebraic Structures of Linear Operators and on Function Spaces (2007)

Vol. 1896: P. Massart, Concentration Inequalities and Model Selection, Ecole d'Été de Probabilités de Saint-Flour XXXIII-2003. Editor: J. Picard (2007)

Vol. 1897: R. Doney, Fluctuation Theory for Lévy Processes, Ecole d'Été de Probabilités de Saint-Flour XXXV-2005. Editor: J. Picard (2007)

Vol. 1898: H.R. Beyer, Beyond Partial Differential Equations, On linear and Quasi-Linear Abstract Hyperbolic Evolution Equations (2007)

Vol. 1899: Séminaire de Probabilités XL. Editors: C. Donati-Martin, M. Émery, A. Rouault, C. Stricker (2007)

Vol. 1900: E. Bolthausen, A. Bovier (Eds.), Spin Glasses (2007)

Vol. 1901: O. Wittenberg, Intersections de deux quadriques et pinceaux de courbes de genre 1, Intersections of Two Quadrics and Pencils of Curves of Genus 1 (2007)

Vol. 1902: A. Isaev, Lectures on the Automorphism Groups of Kobayashi-Hyperbolic Manifolds (2007)

Vol. 1903: G. Kresin, V. Maz'ya, Sharp Real-Part Theorems (2007)

Vol. 1904: P. Giesl, Construction of Global Lyapunov Functions Using Radial Basis Functions (2007)

Vol. 1905: C. Prévôt, M. Röckner, A Concise Course on Stochastic Partial Differential Equations (2007)

Vol. 1906: T. Schuster, The Method of Approximate Inverse: Theory and Applications (2007)

Vol. 1907: M. Rasmussen, Attractivity and Bifurcation for Nonautonomous Dynamical Systems (2007)

Vol. 1908: T.J. Lyons, M. Caruana, T. Lévy, Differential Equations Driven by Rough Paths, Ecole d'Été de Probabilités de Saint-Flour XXXIV-2004 (2007)

Vol. 1909: H. Akiyoshi, M. Sakuma, M. Wada, Y. Yamashita, Punctured Torus Groups and 2-Bridge Knot Groups (I) (2007)

Vol. 1910: V.D. Milman, G. Schechtman (Eds.), Geometric Aspects of Functional Analysis. Israel Seminar 2004-2005 (2007)

Vol. 1911: A. Bressan, D. Serre, M. Williams, K. Zumbrun, Hyperbolic Systems of Balance Laws. Cetraro, Italy 2003. Editor: P. Marcati (2007)

Vol. 1912: V. Berinde, Iterative Approximation of Fixed Points (2007)

Vol. 1913: J.E. Marsden, G. Misiołek, J.-P. Ortega, M. Perlmutter, T.S. Ratiu, Hamiltonian Reduction by Stages (2007)

Vol. 1914: G. Kutyniok, Affine Density in Wavelet Analysis (2007)

Vol. 1915: T. Bıyıkoğlu, J. Leydold, P.F. Stadler, Laplacian Eigenvectors of Graphs. Perron-Frobenius and Faber-Krahn Type Theorems (2007)

Vol. 1916: C. Villani, F. Rezakhanlou, Entropy Methods for the Boltzmann Equation. Editors: F. Golse, S. Olla (2008)

Vol. 1917: I. Veselić, Existence and Regularity Properties of the Integrated Density of States of Random Schrödinger (2008)

Vol. 1918: B. Roberts, R. Schmidt, Local Newforms for GSp(4) (2007)

Vol. 1919: R.A. Carmona, I. Ekeland, A. Kohatsu-Higa, J.-M. Lasry, P.-L. Lions, H. Pham, E. Taflin, Paris-Princeton Lectures on Mathematical Finance 2004. Editors: R.A. Carmona, E. Çinlar, I. Ekeland, E. Jouini, J.A. Scheinkman, N. Touzi (2007)

Vol. 1920: S.N. Evans, Probability and Real Trees. Ecole d'Été de Probabilités de Saint-Flour XXXV-2005 (2008)

Vol. 1921: J.P. Tian, Evolution Algebras and their Applications (2008)

Vol. 1922: A. Friedman (Ed.), Tutorials in Mathematical BioSciences IV. Evolution and Ecology (2008)

Vol. 1923: J.P.N. Bishwal, Parameter Estimation in Stochastic Differential Equations (2008)

Vol. 1924: M. Wilson, Littlewood-Paley Theory and Exponential-Square Integrability (2008)

Vol. 1925: M. du Sautoy, L. Woodward, Zeta Functions of Groups and Rings (2008)

Vol. 1926: L. Barreira, V. Claudia, Stability of Nonautonomous Differential Equations (2008)

Vol. 1927: L. Ambrosio, L. Caffarelli, M.G. Crandall, L.C. Evans, N. Fusco, Calculus of Variations and Non-Linear Partial Differential Equations. Cetraro, Italy 2005. Editors: B. Dacorogna, P. Marcellini (2008)

Vol. 1928: J. Jonsson, Simplicial Complexes of Graphs (2008)

Vol. 1929: Y. Mishura, Stochastic Calculus for Fractional Brownian Motion and Related Processes (2008)

Vol. 1930: J.M. Urbano, The Method of Intrinsic Scaling. A Systematic Approach to Regularity for Degenerate and Singular PDEs (2008)

Vol. 1931: M. Cowling, E. Frenkel, M. Kashiwara, A. Valette, D.A. Vogan, Jr., N.R. Wallach, Representation Theory and Complex Analysis. Venice, Italy 2004. Editors: E.C. Tarabusi, A. D'Agnolo, M. Picardello (2008)

Vol. 1932: A.A. Agrachev, A.S. Morse, E.D. Sontag, H.J. Sussmann, V.I. Utkin, Nonlinear and Optimal Control Theory. Cetraro, Italy 2004. Editors: P. Nistri, G. Stefani (2008)

Vol. 1933: M. Petkovic, Point Estimation of Root Finding Methods (2008)

Vol. 1934: C. Donati-Martin, M. Émery, A. Rouault, C. Stricker (Eds.), Séminaire de Probabilités XLI (2008)

Vol. 1935: A. Unterberger, Alternative Pseudodifferential Analysis (2008)

Vol. 1936: P. Magal, S. Ruan (Eds.), Structured Population Models in Biology and Epidemiology (2008)

Vol. 1937: G. Capriz, P. Giovine, P.M. Mariano (Eds.), Mathematical Models of Granular Matter (2008)

Vol. 1938: D. Auroux, F. Catanese, M. Manetti, P. Seidel, B. Siebert, I. Smith, G. Tian, Symplectic 4-Manifolds and Algebraic Surfaces. Cetraro, Italy 2003. Editors: F. Catanese, G. Tian (2008)

Vol. 1939: D. Boffi, F. Brezzi, L. Demkowicz, R.G. Durán, R.S. Falk, M. Fortin, Mixed Finite Elements, Compatibility Conditions, and Applications. Cetraro, Italy 2006. Editors: D. Boffi, L. Gastaldi (2008)

Vol. 1940: J. Banasiak, V. Capasso, M.A.J. Chaplain, M. Lachowicz, J. Miękisz, Multiscale Problems in the Life Sciences. From Microscopic to Macroscopic. Będlewo, Poland 2006. Editors: V. Capasso, M. Lachowicz (2008)

Vol. 1941: S.M.J. Haran, Arithmetical Investigations. Representation Theory, Orthogonal Polynomials, and Quantum Interpolations (2008)

Vol. 1942: S. Albeverio, F. Flandoli, Y.G. Sinai, SPDE in Hydrodynamic. Recent Progress and Prospects. Cetraro, Italy 2005. Editors: G. Da Prato, M. Röckner (2008)

Vol. 1943: L.L. Bonilla (Ed.), Inverse Problems and Imaging. Martina Franca, Italy 2002 (2008)

Vol. 1944: A. Di Bartolo, G. Falcone, P. Plaumann, K. Strambach, Algebraic Groups and Lie Groups with Few Factors (2008)

Recent Reprints and New Editions

Vol. 1702: J. Ma, J. Yong, Forward-Backward Stochastic Differential Equations and their Applications. 1999 – Corr. 3rd printing (2007)

Vol. 830: J.A. Green, Polynomial Representations of GL_n, with an Appendix on Schensted Correspondence and Littelmann Paths by K. Erdmann, J.A. Green and M. Schoker 1980 – 2nd corr. and augmented edition (2007)

Vol. 1693: S. Simons, From Hahn-Banach to Monotonicity (Minimax and Monotonicity 1998) – 2nd exp. edition (2008)

Vol. 470: R.E. Bowen, Equilibrium States and the Ergodic Theory of Anosov Diffeomorphisms. With a preface by D. Ruelle. Edited by J.-R. Chazottes. 1975 – 2nd rev. edition (2008)

Vol. 523: S.A. Albeverio, R.J. Høegh-Krohn, S. Mazzucchi, Mathematical Theory of Feynman Path Integral. 1976 – 2nd corr. and enlarged edition (2008)